Intellectual Property Rights in Agricultural Research

Intellectual Property Rights in Agricultural Research

Dr Letha Devi G
Senior Scientist (Agri. Extension)
ICAR-NIANP
Adugodi, Bengaluru-560030, Karnataka, India
e-mail: lethaayur@gmail.com

Dr Mukund A. Kataktalware
Senior Scientist (Livestock Production Management)
Dairy Production Section, Southern Regional Station,
ICAR-National Dairy Research Institute (DU),
Adugodi, Bengaluru-560030, Karnataka, India
e-mail: mtalware@gmail.com

Dr Atul P. Kolte
Scientist, ICAR-NIANP, Adugodi
Bengaluru-560030, Karnataka, India

International Books & Periodical Supply Service
(Publisher of Scientific Books)
38, NISHANT KUNJ, PITAM PURA MAIN ROAD
DELHI-110034 (India)
Phone : 011-27352078, Mobile : 09810146811, E-mail : hkjain1975@yahoo.com

Published by :

International Books & Periodical Supply Service

(Publisher of Scientific Books)
38, NISHANT KUNJ, PITAM PURA MAIN ROAD
DELHI-110034 (India)
Phone : 011-27352078, Mobile : 09810146811, E-mail : hkjain1975@yahoo.com

ISBN 978-81-93906-59-0

Composed, Designed & Printed in India

Preface

Farming, in the recent times, has become technology intensive and warrants intellectual property protection to safeguard indigenously developed knowledge and farmer's interest. With the advent of biotechnology and genetic engineering and multilateral agreements, a new era of intellectual property rights (IPR) has begun. The growing impact of technologies in agriculture sector has led to an increasing recognition of the potential of these IPRs as intellectual property (IP) assets. Questions on sharing the monetized benefits of technology products, the intersection areas of the use of genetic resources and associated knowledge in biotechnology based research are posing new challenges to stakeholders in the National Agricultural Research System (NARS). To realize the social and economic benefits of IP rights in agriculture sector, a comprehensive understanding of the key concepts and legal framework on the IP system in this sector is essential. This requires practical insights of how the IP system operates at the ground level, and how these rights can be used and managed effectively for societal benefit.

It is expected that the niche products from Indian agriculture would get remunerative price through IP protection and patenting. As the technological intervention progresses, the role and importance of intellectual property protection also assumes greater importance. To safeguard the interest of all stakeholders of farm sector, it is necessary to protect innovations in agriculture sector through appropriate IP instruments apart from conservation and sustainable use of native varieties, breeds and associated traditional knowledge through *sui generis* system. There are plenty of IP instruments influencing agriculture sector development, access and transfer of genetic resources and some of them are little complex to understand. Scientists and policy-makers are facing challenges in formulating and implementing IP policies influencing research and development in agriculture sector.

Editors

About the Editors

Dr Letha Devi G did her graduation in Agriculture from Kerala Agricultural University in 2001 and topped the university. She completed her Master's in Dairy Extension from National Dairy Research Institute in 2004 and topped the university with Director's Gold Medal. She completed her doctoral degree from Indian Agricultural Research Institute, New Delhi in the discipline of Agricultural Extension in the year 2007. She joined Indian Council of Agricultural Research as a scientist in 2007. She is working as a Scientist under ICAR-National Institute of Animal Nutrition and Physiology, Bangalore. She did PG Diploma in Intellectual Property Rights from IGNOU, New Delhi, in 2013. She has published a number of research papers and popular articles in various national and international journals of repute. She has published a book on livelihood security of rural women in Kerala.

Dr Mukund A Kataktalware did his graduation in Veterinary Sciences and Animal Husbandry from Veterinary College, Nagpur in 1997. He completed her Master's and Doctoral degree in Livestock Production and Management from National Dairy Research Institute in 2001 and 2004 respectively. Dr. Kataktalware joined Indian Council of Agricultural Research as a scientist in 2004 at National Research Centre on Yak, Dirang, Arunachal Pradesh, where he served for about five years. Presently, Dr. Kataktalware is working as a Senior Scientist under ICAR-National Dairy Research Institute, Southern Regional Station, Bangalore. He did his PG Diploma in Intellectual Property Rights from IGNOU, New Delhi in 2013. He has been teaching PG and Ph.D. scholars on IPR management in Agriculture. He has published a number of research papers and popular articles in various national and international journals of repute. He has published a book on Scientific Dairy Farming and successfully guided several PG students for their research work.

Dr Atul P Kolte was born on 15 March, 1974 and earned his bachelor's degree in Veterinary Science and Animal Husbandry in 1996. Later he joined Indian Veterinary Research Institute, Izatnagar for Master's degree in Animal Biotechnology. He joined Agricultural Research Service of ICAR in 2000. Presently, he is pursuing his doctoral program at Jain University, Bangalore. He has experience in handling the intellectual property and technology management services for past one decade.

Contents

Chapter 1

An Overview of Intellectual Property Rights

Letha Devi G, Atul P Kolte, A Dhali, N K S Gowda and Mukund A Kataktalware

Intellectual Property Rights are legal rights, which result from intellectual activity in industrial, scientific, literary & artistic fields. These rights safeguard creators and other producers of intellectual goods & services by granting them certain time-limited rights to control their use. Protected IP rights like other property can be a matter of trade, which can be owned, sold or bought. These are intangible and non-exhausted consumption.

Types/Tools of IPRs

1. Patents
2. Trademarks
3. Copyrights and related rights
4. Geographical Indications
5. Industrial Designs
6. Trade Secrets
7. Layout Design for Integrated Circuits
8. Protection of New Plant Variety.

1. Patents

22A patent is an exclusive right granted for an invention, which is a product or a process that provides a new way of doing something, or offers a new technical solution to a problem. It provides protection for the invention to the owner of the patent. The protection is granted for a limited period, i.e., 20 years. Patent protection means that the invention cannot be commercially made, used, distributed or sold without the patent owner's consent. A patent owner has the right to decide who may-or may not-use the patented invention for the period in which the invention is protected. The patent owner may give permission to, or license, other parties to use the invention on mutually agreed terms. The owner may also sell the right to the invention to someone else, who will then become the new owner of the patent. Once a patent expires, the protection ends, and an invention enters the public domain, that is, the owner no longer holds exclusive rights to the invention, which becomes available to commercial exploitation by others.

All patent owners are obliged, in return for patent protection, to publicly disclose information on their invention in order to enrich the total body of technical knowledge in the world. Such an ever-increasing body of public knowledge promotes further creativity and innovation in others. In this way, patents provide not only protection for the owner but also valuable information and inspiration for future generations of researchers and inventors.

For general principles governing the patent system in India and further details, please refer http://ipindia.nic.in/ipr/patent/patents.htm

2. Trademarks

A trademark is a distinctive sign that identifies certain goods or services as those produced or provided by a specific person or enterprise. It may be one or a combination of words, letters, and numerals. They may consist of drawings, symbols, three- dimensional signs, such as the shape and packaging of goods, audible signs, such as music or vocal sounds, fragrances, or colours used as distinguishing features. It provides protection to the owner of the mark by ensuring the exclusive right to use it to identify goods or services, or to authorize another to use it in return for payment. It helps consumers identify and purchase a product or service because its nature and quality, indicated by its unique trademark, meets their needs. Registration of trademark is prima facie proof of its ownership giving statutory right to the proprietor. Trademark rights may be held in perpetuity. The initial term of registration is for 10 years; thereafter it may be renewed from time to time.

For general principles, governing the trademarks system in India and further details, please refer http://ipindia.nic.in/tmr_new/default.htm

3. Copyrights and Related Rights

A Copyright is a legal term describing rights given to creators for their literary and artistic works. The kinds of works covered by copyright include: literary works such as novels, poems, plays, reference works, newspapers and computer programs; databases; films, musical compositions, and choreography; artistic works such as paintings, drawings, photographs and sculpture; architecture; and advertisements, maps and technical drawings. Copyright subsists in a work by virtue of creation; hence it's not mandatory to register. However, registering a copyright provides evidence that copyright subsists in the work & creator is the owner of the work.

Creators often sell the rights to their works to individuals or companies but able to market the works in return for payment. These payments are often made dependent on the actual use of the work, and are then referred to as royalties. These economic rights have a time limit, (other than photographs) is for life of author plus sixty years after creator's death.

For general principles, governing the copyrights and related rights system in India and further details, please refer http://copyright. gov.in/

4. Geographical Indications (GI)

Geographical Indications are signs used on goods that have a specific geographical origin and possess qualities or a reputation that are due to that place of origin. Agricultural products typically have qualities that derive from their place of production and are influenced by specific local factors, such as climate and soil. They may also highlight specific qualities of a product, which are due to human factors that can be found in the place of origin of the products, such as specific manufacturing skills and traditions.

A geographical indication points to a specific place or region of production that determines the characteristic qualities of the product that originates therein. It is important that the product derives its qualities and reputation from that place. Place of origin may be a village or town, a region or a country. It is an exclusive right given to a particular community, hence the benefits of its registration is shared by all the members of the community. Recently, the GIs of goods like chanderi sarees, kullu shawls, wet grinders from Coimbatore, palakkad matta rice, etc., have been registered.

Keeping in view the large diversity of traditional products spread all over the country, the registration under GI will be very important in future growth of the tribes/communities/skilled artisans associated with developing such products.

For general principles governing the geographical indication system in India and further details, please refer http://ipindia.nic.in/girindia

5. Industrial Designs

Industrial designs refer to creative activity, which result in the ornamental or formal appearance of a product, and design right refers to a novel or original design that is accorded to the proprietor of a validly registered design. Industrial designs are an element of intellectual property. Under the TRIPS Agreement, minimum standards of protection of industrial designs have been provided for. As a developing country, India has already amended its national legislation to provide for these minimal standards.

The essential purpose of design law is to promote and protect the design element of industrial production. It is also intended to promote innovations in industries. The existing legislation of New Designs Act, 2000 serve its purpose well in the rapid changes in technology and international developments. India has also achieved a mature status in the field of industrial designs and in view of globalization of the economy, the present legislation is aligned with the changed technical and commercial scenario and made to conform to international trends in design administration. This replacement Act is also aimed to enact a more detailed classification of design to conform to the international system and to take care of the proliferation of design related activities in various fields.

For general principles governing the industrial design system in India and further details, please refer http://ipindia.nic.in/ipr/design/designs.htm

6. Trade Secrets

It may be confidential business information that provides an enterprise a competitive edge may be considered a trade secret. Usually these are manufacturing or industrial secrets and commercial secrets. These include sales methods, distribution methods, consumer profiles, advertising strategies, lists of suppliers and clients, and manufacturing processes. Contrary to patents, trade secrets are protected without registration.

A trade secret can be protected for an unlimited period of time but a substantial element of secrecy must exist, so that, except by the use of improper means, there would be difficulty in acquiring the information. Considering the vast availability of traditional knowledge in the country, the protection under this will be very crucial in reaping benefits from such type of knowledge. The trade secret, traditional knowledge are also interlinked/associated with the geographical indications.

7. Layout Design for Integrated Circuits

Semi-conductor Integrated Circuit means a product having transistors and other circuitry elements, which are inseparably formed on a semi-conductor material or an insulating material or inside the semi-conductor material and designed to perform an electronic circuitry function.

The aim of the Semi-conductor Integrated Circuits Layout-Design Act, 2000 is to provide protection of Intellectual Property Right (IPR) in the area of Semi-conductor Integrated Circuit Layout Designs and for matters connected therewith or incidental thereto. The main focus of SICLD Act is to provide for routes and mechanism for protection of IPR in Chip Layout Designs created and matters related to it. The SICLD Act empowers the registered proprietor of the layout-design an inherent right to use the layout-design, commercially exploit it and obtain relief in respect of any infringement. The initial term of registration is for 10 years; thereafter it may be renewed from time to time. Department of Information Technology, Ministry of Communications and Information Technology is the administrative ministry looking after its registration and other matters.

For general principles governing the layout design for integrated circuits system in India and further details, please refer http://mit.gov.in/default.aspx?id=322

8. Protection of New Plant Variety

The objective of this act is to recognize the role of farmers as cultivators and conservers and the contribution of traditional, rural and tribal communities to the country's agro biodiversity by rewarding them for their contribution and to stimulate investment for R & D for the development new plant varieties to facilitate the growth of the seed industry.

The Plant Variety Protection and Farmers Rights Act, 2001 was enacted in India to protect the New Plant Variety; the Act has come into force on 30 October, 2005 through Authority. Initially, 12 crop species have been identified for regd. i.e. Rice, Wheat, Maize, Sorghum, Pearl millet, Chickpea, Green gram, Black gram, Lentil, Kidney bean, etc. India has opted for sui-generic system instead of patents for protecting new plant variety. Department, Agriculture and Cooperation is the administrative ministry looking after its registration and other matters.

For general principles governing the protection of new plant variety system in India and further details, please refer http://www.plantauthority.gov.in

Related Links

- www.wipo.int
- http://www.tifac.org.in/do/pfc/pfc

❑❑❑

Chapter 2

Trademarks and their Importance with Special Reference to Dairy Industry in India

Mukund A Kataktalware and Satish Kulkarni

The trademarks is an extremely valuable form of intellectual property rights and is commonly used by organizations, companies, businesses, institutions and others across the world to consolidate their image. The existence of trademarks can be traced as long as 3,000 years ago, when Indian craftsmen used to engrave their unique signatures on their products. During the Middle Ages, the use of signs to distinguish the goods of merchants and manufacturers expanded further. However, with industrialization the trademark started to play a significant role, and has become a critical factor in the era of global trade. The market-driven economy allows competing manufacturers and traders to offer consumers a variety of goods and services in the same category without any apparent differences. Therefore, the consumers need to be given the assistance that will allow them to consider the alternatives and make their choice between the competing goods which otherwise differ in quality, price and other characteristics. Under such scenario, trademarks help the consumers to distinguish the products of different manufacturer's. By facilitating consumers to exercise their choice between the various goods available on the market, trademarks encourage their owners to maintain and improve the quality of the products sold under the trademark, in order to meet consumer expectations. Thus, trademarks incentivise the producer who constantly makes high-quality goods, which in turn encourages economic growth.

Importance for Industry

A trademark is a mark that is lawfully authenticated for use as a unique identity for any business. Trademarks are considered as a brand identifiers. In the legal language, brand names are known as trademarks. It is a symbolism that the whole world recognizes the organization with. Not only does a trademark help in identifying a company, but also it indicates its reputation in the market. A trademark is what differentiates it from its rivals. In the present times, when the markets are expanding at a sweeping pace and competition is increasing, a unique and registered trademark is an absolute necessity. Therefore, every business entity / organization small or big must develop a trademark for itself, which is unique and is not used by any other company. This would ensure that no other company exploits its brand name or reputation. A start-up company does not have much recognition in its business, but might come up with a product that can revolutionize the market. If it does not have a trademark of its own, then its competitors can easily market that product under their trademark. Also, if an established company with a commendable reputation in the market has an unregistered trademark, then another institution can use its trademark and take a shortcut to its reputation and market dominance. Therefore, every institution, old or new, must have a registered trademark of its own. It serves as an important safeguard for the institution in the environment that it operates in. When a name or a brand develops a reputation and is associated with quality goods or services, a trademark assumes importance. It acts as an indicator of quality and can help create an image for the business. As businesses are responsible for anything bearing their trademark, they take more pride in their products. In order to protect its reputation, a company has to work harder to provide quality products and services consistently. If a trademark is used without authorization, an action for infringement of the registered trademark can be filed, with serious consequences to the infringer. Essentially, trademark law was designed to protect consumers. The trademark indicates to the public that there is a connection between the goods or services and the proprietor of the trademark.

Implications for Indian Dairy Industry

Rapid urbanization, increased purchase power parity of middle class improved standard of life, rising awareness about hygienic food, etc., are creating new markets for the milk, and milk products. In order to cater to market demand, a steady supply of quality milk and milk products is essential. There are many regional players who are doing good business. However, in the absence of any pan-India visibility, they cannot expand their business beyond the local market. Therefore, in order to tap readily available national and international market for indigenous milk products, it is essential to develop certain brands. Thus, for established as well as new businesses, a trademark facilitates wider recognition. In the long term, it enhances brand value and helps business to grow many folds.

Types of Trademarks

Trademarks : Any sign, e.g. word, name, logo, label, etc., which can be represented graphically and is capable of distinguishing the goods or services of a given enterprise from its rivals.

Well-known Trademarks : A sign or mark which has become so popular or well known so to the substantial segment of the public which uses such goods or receives such services that the use of such mark in relation to other goods or services would be likely to be taken as indicating a connection in the course of trade or rendering of services between those goods or services and a person using the relation to the first-mentioned goods or services.

Service Marks : Any sign that enables the consumers to distinguish between the different service providers, such as hotels, laundry, airlines, insurance companies, car rental firms, etc.

Collective Marks : Any sign/mark may be owned by an association which itself does not use the collective mark but whose members may use the collective mark subject to compliance with the requirements fixed in the regulations concerning the use of the collective mark.

Certification Marks : A mark capable of distinguishing the goods or services in connection with which it is used in the course of trade which is certified by the proprietor of the mark in respect of origin, material, mode of manufacture of goods or performance of services, quality, accuracy or other characteristics from goods or services not so certified and registrable respect of those goods or services in the name as proprietor of the certification trade mark, of that person e.g. Agmark, ISI

Categories of Signs

Words: company names, surnames, forenames, geographical names and any other words or sets of words, whether invented or not, and slogans, e.g. 'Apple' for computers.

Letters and Numerals: one or more letters, one or more numerals or any combination thereof, e.g. TATA, 4711 cologne.

Devices: fancy devices, drawings, symbols, two dimensional representations of goods or containers.

Coloured Marks: words, devices and any combinations thereof in colour, as well as colour combinations and colour as such.

Three-Dimensional Signs: shape of the goods or their packaging, three-pointed Mercedes star *Audible Signs (Sound Marks):* two typical categories of sound marks can be distinguished, namely those that can be transcribed in musical notes or other symbols and others (e.g. the cry of an animal).

Olfactory Marks (Smell Marks): certain fragrance.

Other (Invisible) Signs: signs that can be recognized by touch.

The ® indicates a registered trademark, whereas **TM** and **SM** indicate unregistrered trademark and service mark respectively. If ®, TM and SM used in text, it should directly follow the mark. These symbols are usually shown as superscript ($^{®}$).

Criteria of Protectability

i) A trademark must be distinguishable among different products

ii) A trademark should not be a misleading character or violates public order or morality

Protection of Trademark Rights

A trademark can be protected on the basis of either registration or use. The Paris Convention places contracting countries under the obligation to provide for a trademark register. Nearly all countries today provide for a trademark register, and full trademark protection is properly secured only by registration. In countries that have traditionally based trademark protection on the use, the registration of a trademark merely confirms the trademark right that has been acquired by use. Consequently, the first user has priority in a trademark dispute, not the one who first registered the trademark.

Registration of Trademarks

In India, an Office of the Registrar of Trademarks has been established for the maintenance of the trademark Registry. Controller General of Patents and Designs is also the Registrar of Trademarks. The Registration of Trademarks is done in Mumbai, where the Trademarks Registry is situated. There are branch offices of the Registry at Kolkata, Delhi, Chennai and Ahmedabad with separate jurisdictions. The procedure for registration is indicated in Fig. 1. The registration of a trademark is for a period of 10 years, which can be renewed indefinitely on the payment of the prescribed renewal fees.

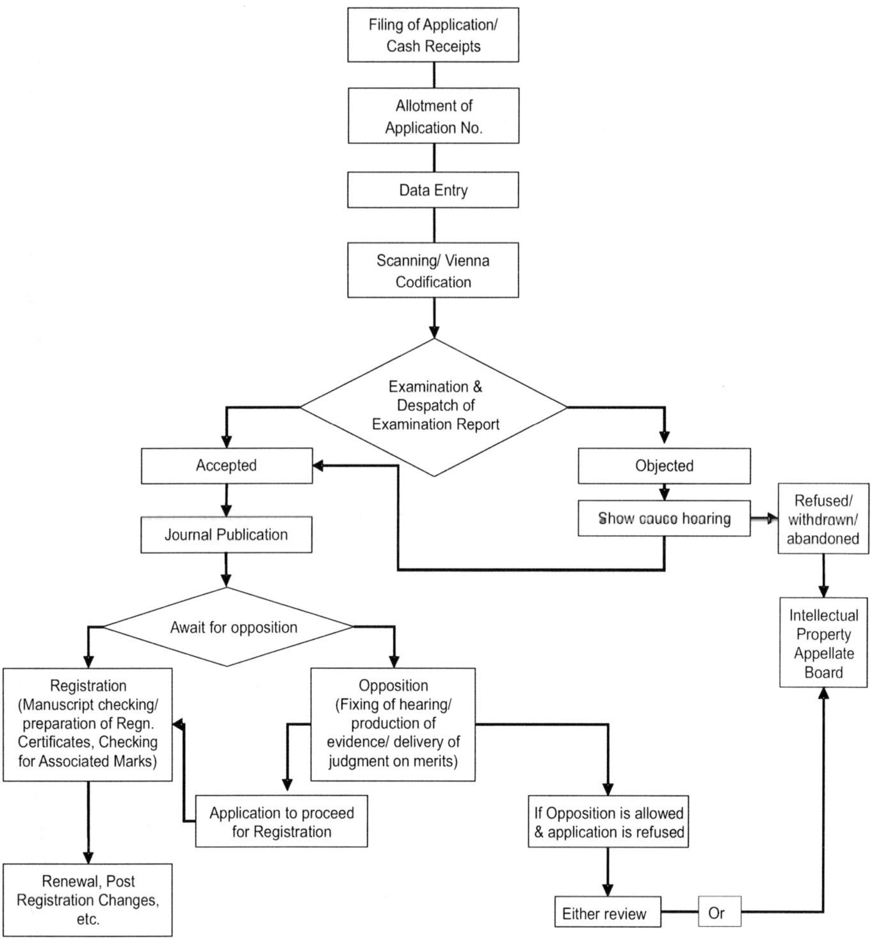

Fig. 1: A flow chart of the process of registering a trademark in India (Adapted from Draft Manual for Trademark Practice and Procedure, 2009)

Rights of Registered Trademarks' Owners

The registration of a trademark serves as proof of adoption and selection of the trademark for specified goods and services. The owner of a registered trademark is entitled to the following rights:

- An exclusive right to the use of the trademark in relation of the trademark to the goods or services in respect of which it has been registered.
- Registration empowers the owner to obtain relief in case of infringement, when a similar mark is used on (a) same, or similar, goods or services, (b) dissimilar goods or services.
- Registration forbids other persons, except the registered, or unregistered permitted user or use the registered trademark or a confusingly similar mark in relation to the same goods or services.
- A registered trademark cannot be used by any one unauthorized in business papers and in advertising.
- Import of goods and services bearing a similar to a registered trademark as a trade name or part of a trade name or name of business concern dealing in the same goods or services.
- A trademark right holder has a right to restrain the use of his registered trademark as a trade name or part of a trade name or name of business concern dealing in the same goods or services.
- Registration grants the right to assign or transmit the trademark to another person by the registered owner of a trademark.

Indian Legislation

In 1940, the then British Government of India passed the Trdemarks Act for uniform and systematic registration of trademarks in India, which came into force on 1st June, 1942. This act was replaced by the Trade and Merchandise Marks Act, 1958, which came into effect from 25th November, 1959. India being the original member of the WTO, it was obligatory for India to amend the Trade and Merchandise Marks Act, 1958 in conformity with the TRIPS. Therefore, the Government of India enacted The Trademark Act, 1999. This Act is in harmony with two major international treaties on the subject, namely The Paris Convention for Protection of Industrial Property and TRIPS Agreement to both of which India is a signatory.

The Trademark Act, 1999 and the Trademarks Rules, 2002 govern the law relating to Trademarks in India. All the provisions of the Trademarks Act, 1999 and the Trademarks Rules came into effect on 15th September, 2003. The legislation, as indicated in the Preamble to the Act, is an "Act to amend and consolidate the law relating to trademarks, to provide for registration and better protection of trade marks for goods and services and for the prevention of the use of fraudulent marks." The important features of this Act are as follows:

- It provides for registration of trademark for services, in addition to goods.
- It does not permit registration of trademarks, which are imitation of well known trademarks.
- Amplifications of factors to be considered for defining a well known mark.
- It provides only a single register with the simplified procedure for registration and with equal rights.
- It simplifies the procedure for registration of the registered user and enlarging the scope of permitted use.
- It enhances punishment for the offences relating to trade marks on par with the Copyright Act, 1957 to prevent the sale of spurious goods.
- There is provision of an Appellate Board for speedy disposal of appeals and rectification applications which at present lie before High Courts.
- It transfers the final authority relating to registration of certification trademarks to the Registrar instead of the Central Government.
- It provides enhanced punishment for the offences relating to trademarks on par with the present Copyright Act, 1957, to prevent the sale of spurious goods.
- It prohibits use of someone else's trademarks as part of corporate names, or name of business concern.
- It extends application of the convention country to include countries which are members of the Group or Union of countries and Inter-governmental Organisations.

- It incorporates other provisions, like amending the definition of "trademarks, provisions for filing a single application for registration in more than one class, increasing the period of registration and renewal from 7 to 10 years; making trade mark offences cognizable, enlarging the jurisdiction of Courts to bring the law in this respect on par with the copyright law, amplifying the powers of the Court to grant *ex parte* injunction in certain cases and other related amendments to simplify and streamline the trademark law and procedure.

Conclusion

Trademarks are important tools of building business/trading image and connote certain amount of confidence in consumers of goods/services. Hence, it is important for any organization to create appropriate trademarks and to prohibit its infringement to protect their business interests. The dairy industry in India is in growth phase and today, there are many brands competing for their business share in the market. Additionally, there is likelihood of products from other countries entering the Indian market, which will further intensify the competition. It is in this context important for Indian dairy industry to develop and maintain trademarks to protect their own interest.

❒❒❒

Chapter 3

Significance of Trade Secrets and Undisclosed Information as Applied to Food Industry

Satish Kulkarni and Subhash S.

Intellectual Property (IP) means the legal rights which arise out of any intellectual activity by an individual or any entity involved in the field of literary, artistic, scientific and industrial, etc. Intellectual property encompasses the tangible representations of intellect and creativity, and is often represented in the form of patents, industrial designs, trademarks, copyright, geographical indications, layout design of integrated circuit rights, and protection of trade secrets.

A trade secret is defined as any set of information that can be used in a business or concern provided that such information is valuable and confidential and can provide actual or potential economic benefits. It is any confidential information with commercial value, reasonably protected from disclosure by its rightful holder. The trade secrets could be a formula, pattern, device or compilation of information which is used in business and which helps to obtain competitive advantage in business over competitors. Further the holder of trade secret takes sufficient measures to preserve the confidential nature (Sri Vidya, 2005).

The Uniform Trade Secret Act, 1970 defines the term trade secret as 'information, including a formula, pattern, compilation, program device, method, technique or process that, (i) derives independent economic value, actual or potential, from neither being generally known to and nor being readily ascertainable by proper means by other persons who can obtain economic value from its disclosure or use and (ii) is the subject of efforts that are reasonable under the circumstances to maintain its secrecy'. A trade secret gives its holder a competitive advantage so long as it remains secret. Trade secrets can include information that is not novel enough to be the subject of patent protection, or not original enough to be protected by copyright (for example, a database of customer lists). The possible reasons for going through the trade secrets systems are: (i) technology may not be patentable but still valuable and needs some kind of legal protection (ii) sometimes a patented invention may need the support of a trade secret for exploitation, (iii) infringement of published technology may in some cases be very difficult to detect and hence, trade secrets are used as a means of exploitation to prevent any such imitation (TIFAC, 1998).

The most obvious purpose of trade secrecy is to maintain order in a competitive market. The trade secret laws are used to prevent 'misappropriations' of the trade secret or to award damages for such misappropriations. The Agreement on Trade related Aspects of Intellectual Property Rights (TRIPS) under WTO, lays down the following three criteria for regarding any information as undisclosed information (or trade secret):

- It must not be generally known or readily accessible by people who normally deal with such type of information,
- It must have commercial value as secret,
- The lawful owner must take reasonable steps to keep it secret.

Significance of Trade Secrets

In today's globalized economy, any individual or an organization/ institution attempts to protect their intellectual property by adopting various available legal measures like patents, copyright, trademark, industrial design, etc. but besides these popular IP rights, there are other IP rights not so popular but which are recently drawing attentions all over the world that is confidential information and trade secret. Confidential information and trade secrets are protected under the common law and there are no statutes that specifically govern the protection of the same. In order to protect trade secrets and confidential information, impermeable agreements should be agreed upon, and they should be supported by sound policies and procedures. For information to be deemed a trade secret, two conditions must be met: first, the information should be confidential, that is to say it cannot be publicly known; second, the trade secret holder must put reasonable efforts to ensure its confidentiality. The factors which determine whether, the information is trade secret includes the extent to which the information is known outside the business, the extent it is known by their employees, the value it has as against their competitors and the time and expense it would take others to duplicate the information.

The importance of trade secrets is increasing greatly as the technology is changing so rapidly and that triggers to have a system to manage the inventions and innovations and protect them in the form of trade secrets. Another significant factor is the relative ease of creating and controlling trade secret rights and the free from the procedural necessities required in case of patents, copy rights, designs, etc. Trade secret rights, in contrast, can be established by the explicit conduct or agreement of the interested parties. A trade secret right starts upon the creation of the idea in some concrete form, and continues as long as secrecy is maintained and trade secrets are volatile form of property, they terminate when secrecy is lost. Nevertheless, trade secrets play a major role in protecting innovations and establishing rights to use new technology.

Examples of Trade Secrets

The following information may be considered as trade secrets:

- Cost and Price Information
- Manufacturing Techniques
- Recipes of food products
- Cost and Pricing data
- Data Compilations
- Computer Programs
- Architectural Plans, etc.

The most famous case for the trade secret, the Coca-Cola formula of the world's famous beverage company. It maintains its secrecy for more than hundred years and still has not entered into the public domain despite the fact that the common ingredients of the drink are known and this secret formula is held as a trade secret among few employees, mostly very senior high level executives. In India, a famous example for trade secret is 'fish medicine' to treat asthama, practiced by Bathini Goud Brothers based in Hyderabad.

Protection of Trade Secrets

Trade secrets are embodied in undisclosed information. The trade secrets are not subject to registration. Unlike other IP rights, the holder of the trade secret need not follow the official procedures to apply to any official authority for protection in order to use the rights arising from the entry into possession of confidential information. Therefore, a trade secret can be protected for an unlimited period of time. As a rule, the protection exists as long as the condition of confidentiality exists. Protection of undisclosed information is the cheapest of all of intellectual property protections and it is a self help approach. In this context, the rightful owner has to bear the only cost for the implementation of the safety and information protection policy, internal control, surveillance, limited third party interaction, and the adoption of measures against trade secret infringers. A trade secret holder can protect himself against unauthorized disclosure and use of the trade secret.

The stipulations of TRIPS in regard to protection of undisclosed information are:

- Natural and legal persons shall have the possibility of preventing information lawfully within their control from being disclosed to, acquired by or used by others without their consent in a manner contrary to honest commercial practices so long as such information is secret; has commercial value and has been subject to reasonable steps to keep the information secret.
- Members, when requiring, as a condition of approving the marketing pharmaceuticals or of agricultural chemicals products which utilise new chemical entities, the submission of undisclosed test or other data, the origination of which involves considerable effort, shall protect such data against unfair commercial use. In addition, members shall protect such data against disclose except where necessary steps are taken to ensure that the data are protected against unfair commercial use (TIFAC, 1998).

Protection of undisclosed information is one form which depends heavily on how well you can guard the technical, commercial and scientific information is being guarded at enterprise level. Sometimes, companies do not protect their inventions through patents for fear of imitation but prefer to maintain the information undisclosed and known only to the concerned persons in the company. The employees are expected to maintain the confidentiality and secrecy of the information. Similarly, in case of technology transfer, the licensee and the licensor are bound by confidentiality agreement for not disclosing the business details to anyone. The concept is equally applicable even during the joint research programmes.

Legal Framework for Protection of Trade Secrets in India

Since trade secrets have not been defined very clearly in most of the countries and many countries do not have a well-defined legislation for handling the cases related to undisclosed information and trade secrets. In India, there is no separate legislation for trade secret protection. Nevertheless, Indian courts have upheld the trade secret protection based on the principles of equity and other legislations in order to protect them. Trade secret owner suffering misappropriations has two major civil remedies. An injunction against further disclosure or use of the trade secret and money damages resulting from loss of the competitive advantage from the secret information. Some of the legislations which are having a connection with the trade secrets are:

- Copyright Act, 1957
- The Designs Act, 2000
- The Information Technology Act, 2000
- The Competition Act, 2002
- The Indian Contract Act
- Indian Penal Code
- Civil Procedure Code
- Criminal Procedure Code

Protection of Confidential Information/Trade Secrets

Since there are no substantive legislations in India to protect trade secrets, the individuals/organizations can pursue their own self help approaches in protecting their secrecy information. Some of the steps that may be taken to protect confidentiality include the following:

- Disclosing the trade secrets strictly on need to know basis.
- Signing of confidentiality agreements i.e. Non-disclosure Agreements by the employees and other concerned in the business.

- Limiting access to the places where the trade secrets could be economically exploited.
- Strong internal processes in an organization like, awareness amongst employees about the secret information, handling of confidential data, security measures for its protection, specifying the authorises personnel to interact with third party disclosures, etc.
- An Exit interview i.e. signing an agreement with employees to return all the information relating to the trade secrets while leaving the employment.

Trade secrets are being adopted in limited way in different fields of specialization to have external monopoly over the invention. However, enactment of strong legislations and adopting stringent organizational procedures and processes for protection of confidential information and trade secrets is essential for securing the competitive advantage in business over competitors for its survival and organizational growth.

References

- Bhupathiraju Raju. (2013). PGD-TMA-421- IP Prosecution and Enforcement. Published by NAARM and UoH, Hyderabad.
- Intellectual Property Rights (IPR) Bulletin, TIFAC, Vol.4 No.3, March, 1998.
- Srividya, G.S. (2005). Nalsar Pro Module. Introduction to IPR. Published by Nalsar University of Law, Hyderabad.
- Verma, S.K. (2002). Legal Protection of Trade Secret and Confidential Information. Journal of Indian Law Institute. Vol. 44: 340.

Chapter 4

Intellectual Property Protection of Microbes and Transgenics

A Thulasi, D Rajendran and M Chandrasekharaiah

The world's population is estimated to reach 9.1 billion by 2050 from the current 7 billion. With the ever increasing population, the key global challenge that would be faced by developing and developed nation would be food security. In view of the impact of climate changes, growing scarcity of water, land and energy security concerns, it has been estimated that we will need to produce 70-100 per cent more food than the current rate without drastically inflating prices to meet the expected demands for food. Crop production methods will also have to sustain the environment and preserve biodiversity. 'Sustainable intensification of agriculture', should be the priority of agricultural research. Crop varieties that produce higher yields and enhance nutritional quality under ever demanding environmental conditions need to be experimented upon. Biotechnology remains one of the most promising tools for achieving sustainable agriculture and genetically modified (GM) crops are at the core of this approach. During the period of 1996 to 2010, the uptake of GM crops increased from 1.7 million hectares to 18 million hectares, i.e., ~87-fold increase, making it the fastest adoption rate for any crop technology transgenic microbes play an impotant role in any field of agriculture - crops, animals and fisheries.

Transgenic Microbes

Bacteria were the first organisms to be modified in the laboratory, due to their simple genetics. These organisms are now used in a variety of tasks, and are particularly important in producing large amounts of pure human proteins for use in medicine e.g. insulin. Genetically, modified bacteria are used to produce the protein insulin, to treat diabetes. Similar bacteria have been used to produce clotting factors to treat haemophilia, and human growth hormone to treat various forms of dwarfism. These recombinant proteins are much safer than the products they replaced, since the older products were purified from cadavers and could transmit diseases. Indeed the human derived proteins caused many cases hepatitis C in haemophiliacs and Creutzfeldt-Jakob disease from human growth hormone. In addition to bacteria being used for producing proteins, genetically modified viruses allow gene therapy. Gene therapy is a relatively new idea in medicine. A virus reproduces by injecting its own genetic material into an existing cell. That cell then follows the instructions in this genetic material and produces more viruses. In medicine, this process is adapted to deliver a gene that could cure disease into human cells. Although gene therapy is still relatively new, it has had some successes. It has been used to treat genetic disorders, such as severe combined immuno-deficiency and treatments are being developed for a range of other incurable diseases, such as cystic fibrosis, sickle cell anaemia, and muscular dystrophy. Tooth decay is caused by *Streptococcus mutans*. This bacterium consumes left over sugars in the mouth and produces acid that eats away tooth enamel and causes cavities. Scientists have recently modified *Streptococcus mutans* to produce ethanol. This transgenic bacterium, if properly colonized in a person's mouth, could possibly eliminate cavities and other tooth related issues. Transgenic microbes have also been used in recent research to kill or hinder tumours, and fight Cohn's disease. Genetically, modified bacteria are also used in some soils to facilitate crop growth, and can also produce chemicals which are toxic to crop pests.

Genetic Modification of Bacteria

Bacteria-the first organisms to be genetically engineered - are used for replicating and altering genes that are subsequently introduced into plants or animals. Bacterial systems lend themselves to genetic manipulation in part because of their rapid reproduction rates. It is easy to produce a genetically identical population - a clone of bacteria - all containing the gene of interest in a short period. The cells can then be lysed and DNA can be isolated in short order. Bacteria are routinely used to produce non-bacterial proteins. An example is the production of purified proteins for vaccine use. Such proteins can be safer than and as effective as vaccines that contain killed or attenuated (weakened) pathogens. Genetic engineering can also produce extensive changes in the bacterium's metabolism. For example, bacteria can be provided with several genes, encoding enzymes that allow the production of fuel alcohol from wood. Researchers have taken advantage of nature to modify bacteria. Plasmids are small, circular, self-replicating, extra chromosomal pieces of DNA that occur naturally. A plasmid can encode a protein that offers its host a selective advantage. For example, a plasmid that encodes an antibiotic allows its host bacterium to thwart competing microbes. Alternately, a bacterium might possess a plasmid that encodes antibiotic resistance. Plasmids are readily isolated from bacterial cells and can be altered *in vitro* by inserting or deleting specific sequences of DNA. Because they can be used to create clones of genes, plasmids are called cloning vectors.

Intellectual Property Rights, Biotechnology and Associated Legal Issues

Biotechnology draws on the pure biological sciences-genetics, microbiology, animal cell culture, molecular biology, biochemistry, embryology, cell biology) and in many instances is also dependent on knowledge and methods from outside the sphere of biology (chemical engineering, bioprocess engineering, information technology and bio-robotics). Conversely, modern biological sciences (including even concepts such as molecular ecology) are intimately entwined and dependent on the methods developed through biotechnology and what is commonly thought of as the life sciences industry. The Biotechnology Sector is one of the fastest growing sectors in India with almost all global majors and large Indian players having established a presence in this sector in India. International players are leveraging the competitively priced but highly qualified and skilled human capital available in India for undertaking Research and Development in India to develop new technology and products, both for Indian and global markets. India further offers a well-developed legal/regulatory system enabling the creation and enforcement of intellectual property and related technology transfer and licensing arrangements.

Strong intellectual property right (IPR) protection is vital for the development of a vibrant biotechnology sector. Development of transgenic (GMOs) microbes requires huge investments in research and development and IPR provides economic incentives to innovators by enabling them to claim benefits associated with it. IPR protection, however, does not guarantee full appropriation of the benefits from the innovation because of expensive monitoring, enforcement and apathy in some cases. Infringement of IPR is, therefore, pervasive. Widespread violation of IPRs especially in developing countries is becoming a contentious issue worldwide. It has been argued that IPR infringements not only hurt both farmers and innovators, but also negatively affect pricing and therefore, represent one of the biggest bottlenecks for adoption of new technologies in developing countries.

Genetic resources and their usage form part of humans' daily existence, from the food ingested to the medicine used to cure ailments to the advancement of standards of human health and well-being. Scientific research and commercial application, particularly with regard to the genetic manipulation of DNA, have permitted these advancements to be made. However, the legal constructions currently in operation are not appropriate to fully recognize these with the necessary certainty both in the short and long term, particularly if increased disclosure obligations are required or the requirements of novelty or inventiveness are no longer compelled to be components of proprietary protection. If there is a requirement of lot of disclosure, then this would introduce legal uncertainty to the system and to the trade in such resources and the commodities that they generate. Coupled with these scientific advancements or achievements are the issues of associated traditional knowledge concerning genetic resources and their subsequent use, particularly for the purpose of commercialization creating problems. These technically cultural concerns do not fit within the legal paradigm currently in operation at an international level and raise questions relating to the equity of existing arrangements. However, a large proportion of traditional knowledge is currently within the public domain and to offer legal protection to this knowledge would create legal uncertainty. The issue, of which traditional knowledge requires protection, public or private domain, requires clarification. If international legal protection is granted to the body of traditional knowledge currently held within the public domain, the resultant effect would be a creation of a level of legal uncertainty regarding the exchange and use of genetic resources for both, research and commercial purposes. Staying at the forefront of science with cutting-edge innovations and the need to protect such innovations has gained importance since the foundation of the World Trade Organization (WTO). Enriched by the tools of molecular biology, bio-processing has a pivotal role in the production of biochemicals, bioenergy, biomedicines and bioremediation. In recent years, newer bio-industrial processes have been developed to tap the vast, invisible and still-to-be-explored reservoir of microbial diversity. Microbial resources used for these applications are preserved in microbial culture collections, which are essential for supporting scientiûc development both, in the developed and the developing world. The Uruguay Round Agreement on trade-related aspects of intellectual property rights included microbial patenting as a component and countries that have signed the agreement have to allow microbial patenting from the year 2005 onwards, if it is not pre-existing. Patent law requires disclosure of the full details of an invention. Descriptions and drawings are inadequate and insufficient, if the invention involves the use of micro-organisms or microbial components. Therefore, it has been agreed internationally that microbial samples must be deposited in a culture collection that is recognized as an 'international depository authority' (IDA) within the framework of the Budapest Treaty on the international recognition of the deposit of micro-organisms for the purposes of patent procedure of the 146 signatories to WTO, only 20 countries have a culture collection with IDA status (as of 4 April, 2003). Thus, it is significant to analyze the modalities required for getting IDA status, as well as to identify the problems that will exist in countries without IDA status. Moreover, in accordance with the Uruguay Round TRIPS Agreement, all of the developing countries must implement both, microbial and product patenting from the year 2005 onwards. It is high time that developing countries discuss these implications and evolve strategies for effective implementation. In this review, we discuss the ways and means of meeting the requirements for acquiring IDA status.

Defining micro-organisms: an ongoing paradox the use of the term 'micro-organisms' has created considerable confusion. Generally, microbes or micro-organisms are tiny living things that are invisible to the naked eye. For the purposes of patent protection, the term micro-organism often applies to biological material including, for example, viruses, bacteria, actinomycetes, yeasts, filamentous fungi, mushrooms, protozoa, unicellular algae, cell lines of plants or animals, fused cells, transformants and vectors used in genetic engineering, variants, plant cells, DNA and RNA. In many countries, there is uncertainty regarding the scope of the term 'micro-organism'; furthermore, it is not possible to confine the term to only living things because culture collections deal with diverse biological materials. In fact, the European Union has decided to discontinue the use of the term 'micro-organism'; instead, it has decided to use the term 'biological material', which means any material that contains genetic information and is capable of replicating itself or being reproduced in a biological system.

Challenges in deposition and the depositor's role although bacteria, viruses, fungi, plant cells, animal cells and so on differ marginally in the method and manner of their cultivation and storage, the basic facilities that are required are the same. Thus, culture collections can deal with various biological materials. At present, it is under-stood that IDAs are intended for storing any microscopic life form or microbiota (microscopic flora and fauna). It is important that IDAs should demarcate naturally isolated micro-organisms from genetically modified micro- organisms to preserve the natural gene pool. Another challenge is how to preserve the natural gene pool without the accumulation of spontaneous mutations and thereby to maintain the originality of the strains deposited. In addition, IDAs should be prepared to deal with the emerging need of microbial consortia (natural or artificial assemblages of interacting micro-organisms), microcosms (natural or artificial sites that not only are defined completely in chemical and physical properties but are also gnotobiotics, that is, composed of known species that occupy ecological niches), and viable but not yet culturable micro-organisms that might ultimately lead to the preservation of microbes that mimic the natural ecosystems. Such properties will pave the way for the transformation of IDAs into in situ culture reserves (places where biodiversity is maintained in nature, for example, ecological reserves, natural habitats and farms). Notably, 99% of the existing microbial diversity has not yet been explored [4] and thus, microbial handling, preservation and operation of IDAs will become more challenging. It is also significant that IDAs need to store non-living biological materials, such as oncogenes, plasmids, RNA and eukaryotic DNA that are replicable in biological hosts. They therefore need to acquire expertise in handling these materials. Owing to the continuous developments in the fields of rDNA, whole-genome sequencing and bio-informatics, dealing with these non- living biomaterials is an additional challenge. The various organisms accepted by the reputed repositories is summarised in Table 1.

Table 1. Types of 'micro-organism' accepted by some of the principal international depository authority

Country	International depository authority	Micro-organisms accepted
Belgium	Belgium Coordinated Collections of Micro-organisms (BCCM)	Animal cell cultures, bacteria (pathogenic and non- pathogenic), fungi (pathogenic and non-pathogenic), human cell cultures, hybridomas, oncogenes, plasmids (In hosts and not in hosts), RNA, yeasts (pathogenic and non-pathogenic)
Germany	Deutsche Sammlung von Micro-organism en und Zellkulturen GmbH (DSMZ)	Animal cell cultures, bacteria(non-pathogenic), bacteriophages, fungi (non-pathogenic), human cell cultures, murine embryos, plant cell cultures, plant viruses, plasmids (in hosts and not in hosts), yeasts (non-pathogenic)
Japan	International Patent Organism Depositary (IPOD)	Algae, animal cell cultures, non-pathogenic bacteria, embryos, non-pathogenic fungi, plant cell cultures, plasmids (in hosts and not in hosts) non-parasitic protozoa, seeds, and non-pathogenic yeasts
UK	Culture Collection of Algae and Protozoa (CCAP) CABI Bioscience, UK European Collection of Cell Cultures (ECACC) Centre Non-pathogenic bacteria, non-pathogenic fungi, non-pathogenic yeast National Collection of Type Cultures (NCTC)	Algae, non-parasitic protozoa Pathogenic bacteria and plasmids (in hosts) Animal viruses, animal cell cultures, bacteria (pathogenic and non-pathogenic), eukaryotic DNA, hybridomas, plant cell cultures, protozoa (parasitic and pathogenic), pathogenic yeasts
USA	Agricultural Research Service Culture Collection (Northern Regional Research Laboratory) American Type Culture Collection (ATCC) National Collection of Yeast Cultures (NCYC) National Collections of Industrial, Food and Marine Bacteria (NCIMB)	Bacteria (non-pathogenic), fungi (non-pathogenic), molds, plasmids (in hosts), non-pathogenic yeasts. Algae, animal viruses, animal cell cultures, bacteria (pathogenic and non-pathogenic), bacteriophages, embryos, eukaryotic DNA, fungi (pathogenic and non-pathogenic), human cell cultures, hybridomas, molds, murine embryos, mycoplasma, oncogenes, plant cell cultures, plant viruses, plasmids (in hosts and not in hosts), protozoa (parasitic, non- parasitic and pathogenic), RNA, seeds, yeasts (pathogenic and non-pathogenic) Plasmids (in hosts) and non-pathogenic yeasts

Future of IDAs

In the near future, IDAs might be replaced by 'biological resource centres (BRCs); such centres will be the ultimate repositories because they aim to function also as information banks by compiling databases of molecular, physiological and structural information relevant to the collections, as well as related to bio-information. The concept of BRCs came from worldwide work at the Organization for Economic Co-operation and Development; as discussed at the Tokyo Workshop on Biological Resource Centres, 1999: "BRCs are an essential part of the infrastructure underpinning life sciences and biotechnology". They consist of service providers and repositories of the living cells, genomes of organism and information relating to heredity and functions of biological systems. BRCs contain collections of culturable organisms (e.g. micro-organisms, plant, animal and human cells), replicable parts of these (e.g. genomes, plasmids, viruses, cDNAs), viable but not yet culturable organisms, cells and tissues, as well as databases containing molecular, physiological and structural information relevant to these collections and related, bio-informatics" [4]. BRCs must meet the high standards of quality and expertise demanded by the international community of scientists and industry for the delivery of biological information and materials. They must provide access to biological resources on which research and development in life sciences and the advancement of biotechnology depend. France has constituted a consultative committee on BRCs. Currently, there are four important issues under discussion by the committee:

i) Scientific research on networks of genes involved in tissue and cell functioning and malfunctioning requires biological resources of guaranteed origin and quality;

ii) The diversity of biological collections and their uncontrolled emergence entail risks for health and for the environment (e.g. dissemination of pathogenic agents) and cause safety and security concerns;

iii) Although there are legislative and regulatory frameworks for the scientiûc and economic use of biological collections, there are enforcement problems that have not yet been solved;

iv) The uncontrolled trade in biological samples, and the genetic history that is present in these materials, is in danger of being beyond any attempt at regulation. Biological collections are the key to post genomic development and the industrial exploitation of scientific results. There are no measures to ensure that trade in this field is open and visible. In addition, the committee is studying the processes for assessing and accrediting BRCs to form a 'national network'. BRCs should also develop capacities to foster solutions to the issue of intellectual property rights, which will be needed for individualistic depositors and collections.

Importance of Micro-organisms

Given the vital importance of biotechnology to the future of the Indian economy, any policy decisions on biological resources, especially Intellectual Property Rights, must be very carefully thought through. Micro-organisms which play an important role in biotechnology are crucial to modern agriculture, and industry. Given their role in bio-remediation, in the production of alternate sources of fuel, such as biogas, ethanol and solar energy and in enzyme technology, their unrestricted use is vital to ensure the high level production base needed for national development. Any possible patent regime must be discussed in depth before we take a hasty, uninformed decision.

Current Status

Indian institutions and scientists are poorly prepared to function in a microbe patent regime. There is widespread patent illiteracy at all levels in the country. There is a high level of ignorance about drafting and filing for patents among scientists. CSIR and ICAR are not providing any technical assistance. It is, therefore, of vital importance that scientists, attorneys, judges, government functionaries and patent office staff are trained in technical matters relating to patents if we are to meet the challenge posed by the new regime.

Since it is not very clearly understood as to how the new patent regime will affect our agriculture and industry, many more discussions and brain-storming sessions must be held before we take a stand on the matter. Much greater technical understanding is needed, for even simple things like what constitutes a patentable organism.

Action Needed

There is an urgent need to document and characterize our micro flora. India is rich in microbes. Not even 1 % has been identified so far. We need to characterize our micro-organisms to lay claim to our resources and know how to use them. The government has to invest in this area to boost the fields of biosystematics and taxonomy. There has to be better coordination between various government departments and universities which are involved in this field, so that technical skills can be maximized effectively.

The aim of posterity and global betterment through the deposition of micro-organism patents thus encounters several limitations. It is time to maximize efforts to increase the number of IDAs and to augment preservation of the natural gene pool. The need for IDAs and the pertaining procedures should be familiarized among the beneficiaries. It is important to intensify research to solve the problems associated with culture preservation. At present, it is difficult to imagine all of the problems that might arise over the long term during the operation of IDAs, and hence, the discussion of potential issues by an expert committee at a global level is needed. It would be appropriate to include developing countries in the decision-making process because they are also principal stakeholders in biological resources. International co-operation among the culture collections must also be fostered. Finally, we should move towards creating a universally acceptable type of BRC and global common genetic resources.

Chapter 5

Intellectual Property Protection of Innovations in Animal Biotechnology

A Dhali, A P Kolte, Letha Devi G, A Mech, M Sridhar

Introduction

Biotechnology exploits biological materials, living or non-living, and is broadly classified as classical and modern biotechnology. The concept of biotechnology is not new, rather it has existed since ancient times and its progress has been witnessed through many centuries across the world. The Hungarian animal entrepreneur Károly Ereky coined the term "biotechnology" during 1919 to describe a technology based on converting raw materials into a more useful product. Currently, the area includes the old biotechnological processes for producing fermented products as well as the new generation biotechnological interventions through genetic engineering, recombinant DNA technology, methods of isolation of micro-organisms from culture medium, methods of mutation, cultures, mutants, transformants, plasmids, processes for making monoclonal antibodies, cell lines for making monoclonal antibodies and many more. Ever since a genetically engineered micro-organism was granted patent, the area has gained enormous significance in protecting genetically engineered organism and other intellectual assets.

The history of patent law in biotechnology dates back to 500 BCE and evidences indicate that some form of patent rights was recognized in Ancient Greece during that time. Patents were systematically granted in Venice as of 1450, where they issued a decree by which new and inventive devices had to be communicated to the Republic in order to obtain legal protection against potential infringers. The period of protection was 10 years. These were mostly in the field of glass making. Subsequently, as Venetian people emigrated to different countries, they demanded similar intellectual property protection in their new homes. This led to the diffusion of patent systems into other countries. The venetian patent system is considered as the foundation of the current day's patent system in different countries. The requirements of the Venetian patent law, such as utility, novelty and non-obviousness are still the basis of modern patent law around the world.

There are some critical issues relating to patentability of biotechnological inventions, such as novelty, obviousness, industrial applicability, extent of disclosure and clarity in claims. In addition, a few special issues have also evolved, such as those relating to moral and ethical concerns, environmental safety, issues relating to patenting of ESTs (Expressed Sequence Tags) of partial gene sequences, cloning of farm animals, stem cells, gene diagnostics, etc. Therefore, patenting the inventions in the field of biotechnology is challenging task both for the applicants as well as patent office.

Evolution of Indian Patent Act

The first ever patent law in India was introduced by the British rulers, the Act VI of 1856 on protection of inventions based on the British patent law of 1852. After several modifications in 1859, 1872 and 1888, the consolidated Indian Patents and Designs Act, 1911 was implemented and offered patent protection to the inventors. Following independence, with the changing socio-politico scenarios, efforts were made to consolidate the Indian patent law by bringing up a new legislation and the Indian Patent Act, 1970 was implemented. The Patent Act, 1970 highlights the invention that satisfies the universally accepted requirements, novelty, inventive step and industrial application. The Patent Act, 1970 was amended subsequently several times in 1999, 2002 and 2005. The Patent (Amendment) Act, 2005 has introduced product patents on substances intended to be used as medicine, drug or food following the implementation of TRIPS required amendments to its Act.

Considerable changes were made in the patenting procedure through the introduction of Patent Rules, 2003 further amended in 2005 and 2006, resulting in new practice and procedure. The Patent Act does not specify which are patentable, but it illustrates subject matters that are not patentable. The Patent Act, 1970 does not mention anything about biotechnology invention. However, India's patent third laws were amended in 2002 to explicitly include biochemical, biotechnological, and microbiological processes within the definition of potentially patentable chemical processes. Until 2002, patents were not granted for the inventions related to living entities of natural or artificial origin, biological materials or other materials having replicating properties, substances derived from such materials, and any processes for the production of living substances/entities including nucleic acids. In contrast, patents could be granted for the processes of producing non-living substances by chemical processes, bioconversion and microbiological processes using micro-organisms or biological materials. In 2002, as per the decision of the Calcutta High Court, granting patent was allowed to the inventions, where the final product of the claimed process contained living micro-organisms. The court concluded that a new and useful art or process is an invention, and where the end product (even if it contains living organism) is a new article, the process leading to its manufacture is an invention. Subsequently, in the year 2002, the Patent Act, 1970 of India was amended by the Patent (Amendment) Act, 2002 where biochemical, biotechnological and microbiological processes were included within the scope of chemical processes for the grant of patents.

Scope of Biotechnology Patenting in India

Indian law with regards to patenting of biological materials are relatively new and not well-settled. There are several deficiencies which are required to be addressed. Apart from narrow standards of patentability, biotech products are faced with certain additional hurdles like mandatory disclosure of biological material, prior approval from the Biodiversity Board and access and benefit issues under the Indian Patent regime.

Trade-Related Aspects of Intellectual Property Rights (TRIPs) Agreement was one of the most contentious issues in the Uruguay Round of multilateral trade negotiations, which was concluded in 1994. The commitments under the TRIPs Agreement have compelled India to amend its patent regime in 1999, 2002 and 2003. The viewpoint of India's Patent Act of 1970 varies considerably from the framework under TRIPs. There are several knowledge and information areas which are not patentable in India. TRIPs does not allow exclusion of inventions that are novel, non-obvious and industrially applicable, but the Indian law has a long list of exceptions.

Biotechnology inventions patentable in India

- Genetically modified organisms, such as bacteria, fungi, plant and animal.
- Microbiological processes and products from genetically engineered micro-organisms. or the genetically engineered micro-organisms themselves, if they are non-living.
- mRNA which is encoded to express protein.
- cDNA matching the sequence of mRNA which is encoded to express protein.
- Isolated and purified DNA sequence, such as genomic DNA coding for a gene or a fragment thereof.
- Oligonucleotides.
- Proteins and polypeptides.
- DNA markers.
- Recombinant DNA including recombinant plasmids and recombinant vectors.
- Selectable markers and promoters in a vector.
- Regeneration and transformation procedures.
- Agrobacterium strains used for transformation.
- Expressed sequence tag (ESTs), if it has a use
- Hybriodoma technology
- Cell lines, if produced artificially.

Biotechnology inventions not patentable in India

- Invention are not patentable, if it is immoral or against public order, harmful to human, animal or plant life or harmful to environment.
- Discovery of living things or non-living substances in nature.
- Plants and animals in whole or any parts thereof including seeds, varieties and species.
- Essential biological processes for the production or propagation of plants and animals.
- Any process for the medicinal, surgical, curative, prophylactic, diagnostic or therapeutic or other treatment of human beings or animals to render them free of disease or to increase their economic value or that of their products.

- New use or new property of known substance.
- Methods of agriculture or horticulture.
- Traditional knowledge.

Conservation of Biological Diversity and Patent Law

India has implemented the Biological Diversity Act to address the issues of prior informed consent, disclosure and access and benefit sharing. The Biological Diversity Act, 2002 provides mechanism for conservation of biological diversity, sustainable use of its components and fair and equitable sharing of the benefits arising out of the use of biological resources, knowledge and for matters connected therewith or incidental thereto. In order to prevent misappropriation of biological resources of India, the Biological Diversity Act requires that access to the biological resources of India is subject to the equitable benefit sharing through the approval of National Biodiversity Authority (NBA). No Intellectual Property Rights including patents based on research or information on biological resources obtained from India shall be granted without the approval of the NBA.

The Patent Act provides interfaces with the process of obtaining patents and access to and benefits sharing from utilization of Indian biological resources. Thus, disclosure of the source and geographical origin of a biological material used in an application for a patent has been made mandatory.

Patent Law and Traditional Knowledge

Currently Indian Patent Act includes several provisions aimed at traditional knowledge outside the realm of patenting. It excludes patenting an invention which, in effect, is traditional knowledge or which is an aggregation or duplication of known properties of traditionally known component or components. The act also prevents to exercise proprietary rights on India's genetic resources and indigenous knowledge. The Act's disclosure requirements mandate inclusion of the source and geographical origin of biological material used in the claimed invention, and interested parties may oppose or petition to revoke an Indian patent on the ground that the invention claimed therein is anticipated "having regard to the knowledge, oral or otherwise, available within any local or indigenous community in India or elsewhere".

The government of India has granted the European Patent Office and United States Patent and Trademark Office the access to its Traditional Knowledge Digital Library (TKDL). The TKDL online database contains translations of manuscripts and textbooks in five languages (500-5000 year old texts on Ayurvedic cure and herbal remedies are documented) including English. Access to the 30-million-page database helps to correctly examine patent applications relating to traditional knowledge. With the TKDL, examiners have improved access to background information at an early stage of patent examination. Prominent cases of patent disputes include a US patent on the wound-healing properties of turmeric (revoked in 1997) as well as an anti-fungal product from the Indian Neem tree (revoked in 2008). Both herbal practices were evidence of traditional knowledge and the patent were revoked.

Conclusion

Patenting the biotechnological inventions in India is relatively new. Well-settled guidelines and legislations are yet to be established relating to patenting or non-patenting of biotechnological inventions. Usually, if the claim of an invention is novel and inventive and modified genetic material, which has an industrial application, Patent Office generally accepts such invention as patentable. The patenting of biological material in India is still decided more often on a case-by-case basis. India's post-TRIPS patent laws include several provisions that make biotechnology patenting more attractive as a basis for competitive advantage.

References

- Guidelines for examination of biotechnology applications for patent. Office of the Controller General of Patents, Designs and Trade Marks, Govt. of India, March, 2013. pp 1-22.
- Manual of patent office practice and procedure, Office of The Controller General of Patents, Designs and Trade Marks, Govt. of India, March, 2011. pp 1/1-25/8.
- Biotechnology patenting In India and related issues. Compiled by Rajashree Sharma, Corporate Law Group, New Delhi, India.

❐❐❐

Chapter 6

TRIPS Agreement and Protection of Innovations in the Field of Biological Science

P Krishnamoorthy

The agreement on **Trade-Related Aspects of Intellectual Property Rights** (**TRIPS**) is an agreement administered by the World Trade Organization (WTO) that sets down minimum standards for many forms of Intellectual Property (IP) regulation as applied to nationals of WTO Members. It was negotiated at the end of the Uruguay Round of the General Agreement on Tariffs and Trade (GATT) in 1994. At present, the products of intellectual creativity, such as inventions, designs, know-how and artistic works serve an important role, and in order to promote such creative activity, inventions, industrial designs, literary and artistic works, layout-designs of integrated circuits, trade secrets and so on are given protection. Furthermore, trademarks and other such signs are also protected so as to secure the trust obtained through business activities, as well as protects consumers and ensure fair competition. In International trade, the proportion of goods and services consisting of Intellectual Property has increased dramatically, and if the intellectual property protection offered by countries is inadequate or inappropriate, there is a danger of distorting the international trade order. However, in developing countries, although systems existed for the protection of intellectual property, there were many countries where the standard of protection was inadequate, for example, the extent of protection was limited or the period of protection was extremely short, or enforcement of intellectual property rights could not be guaranteed to be sufficiently effective. Even among developed nations, there were some countries with systems that gave too much protection to intellectual property or discriminated between internal and external sources.

In this situation, the negotiations concerning Trade-Related Aspects of Intellectual Property Rights (TRIPS) became one of the important new areas for discussion at the Uruguay Round of GATT, begun in 1986. Along with other Agreements to come out of the Uruguay Round, the Agreement on Trade-Related Aspects of Intellectual Property Rights was finally agreed upon at the ministerial meeting in Marrakesh, Morocco in April 1994, and came into under the TRIPS agreement member countries are obliged to provide protection to plant varieties either by patents or by effective *sui generis* system or by any combination thereof (Article 27.3(b)). According to Article 27 of the Agreement, patents have to be available for any inventions, whether products or processes, in all fields of technology. This includes biotechnology. Article 27.3 is a derogation clause in accordance with which members may exclude from patentability plants and animals other than micro-organisms, and essentially biological processes for the production of plants or animals other than non-biological and microbiological processes. Frequently, discussions on the TRIPs Agreement as it relates to biological diversity and the interests of indigenous peoples and local communities focus on patents, in particular the highly emotive issue of patenting life and the availability of the so-called '*sui generis*' option for plant varieties (Article 27[3b]). Without denying the importance of patents, one must bear in mind that the Agreement's 73 articles provide a variety of intellectual property rights.

Key points

- The rules under the Agreement allow countries not to patent higher organisms, whether plants or animals, and essentially biological processes for the production of plants and animals.
- Members must generally provide patent-protection for micro-organisms and for non-biological and micro-biological processes.
- Countries must also protect plant varieties by patents, by an effective *sui generis* system or by any combination thereof.
- The patent provisions under the TRIPS Agreement are not necessarily appropriate to the protection of living materials or products related to there. A *sui generis* system may offer more flexibility in designing a legal means of protection.

- Considerable scope for interpretation remains, when considering the obligations imposed by the Agreement due to the absence of definitions for key terms used in the Agreement.

Requirements of TRIPS

TRIPS require member states to provide strong protection for intellectual property rights. For example, under TRIPS:

- Patents must be granted for "inventions" in all "fields of technology" provided they meet all other patentability requirements (although exceptions for certain public interests are allowed (Art. 27.2 and 27.3) and must be enforceable for at least 20 years (Art 33).
- Copyright terms must extend at least 20 years, unless based on the life of the author. (Art. 12 and 14)
- Copyright must be granted automatically, and not based upon any "formality," such as registrations, as specified in the Berne Convention. (Art. 9)
- Computer programs must be regarded as "literary works" under copyright law and receive the same terms of protection.
- National exceptions to copyright (such as "fair use" in the United States) are constrained by the Berne three-step test.
- Exceptions to exclusive rights must be limited, provided that a normal exploitation of the work (Art. 13) and normal exploitation of the patent (Art 30) is not in conflict.
- No unreasonable prejudice to the legitimate interests of the right holders of computer programs and patents is allowed.
- Legitimate interests of third parties have to be taken into account by patent rights (Art 30).
- In each state, intellectual property laws may not offer any benefits to local citizens which are not available to citizens of other TRIPS signatories under the principle of national treatment (with certain limited exceptions, Art. 3 and 5). TRIPS also has a most favoured nation clause.

Many of the TRIPS provisions on copyright were copied from the Berne Convention for the Protection of Literary and Artistic Works and many of its trademark and patent provisions were modelled on the Paris Convention for the Protection of Industrial Property.

Major Characteristics of the TRIPS Agreement

Objectives

Article 7, establishing the objectives of the TRIPS Agreement, states that "The protection and enforcement of intellectual property rights should contribute to the promotion of technological innovation and to the transfer and dissemination of technology".

Minimum standards

The TRIPS Agreement establishes minimum standards for the protection of Intellectual Property. States are free to introduce systems of protection not referred to in the Agreement, for example, to protect informal knowledge, or the rights of indigenous communities. They are also free to set higher standards of protection for existing rights.

Enforcement

The Agreement establishes general principles that are applicable to IPR Enforcement procedures so that rights' holders can effectively enforce their rights. In addition, it contains provisions on court procedures, court orders, suspension of custom clearances for goods and criminal penalties.

Dispute settlement

The WTO includes a procedure for settling disputes between Members. Any Member can bring proceedings against another Member. A panel of specially appointed trade experts interprets the various agreements of the WTO, including the TRIPS Agreement. Once the panel issues its report, it is adopted unless one of the parties to the dispute appeals the decision or there is a consensus by WTO Members not to adopt the decision. If a party to a dispute fails to abide by a decision of either a Panel or the Appellate Body, the other party can impose trade sanctions on that Member.

Built-in agenda

The TRIPS Agreement includes a built-in agenda of review. Specifically, Article 27(3)(b) provides for the review in 1999 of certain provisions relating to the patentability of plants and animals, and the protection of plant varieties. In the area of geographical indications,

Article 23(4) provides that, in order to facilitate the protection of geographical indications for wines and spirits, negotiations shall be undertaken in the TRIPS Council on the establishment of a multilateral system of notification and registration. Preliminary work in this area has already begun. Article 71(1) provides that the TRIPS Council shall review implementation of the whole Agreement from 1 January, 2000 and every two years after that.

Developing countries' transitional periods

While developed countries should have complied with the obligations under the TRIPS Agreement by 1 January, 1996, developing countries and countries with economies in transition have until 1 January, 2000. Least developed countries have given until 2006 to implement the Agreement.

Sufficient disclosure and best mode

TRIPS agreement provides that, "the applicant for patents shall disclose the invention in a manner sufficiently clear and complete for the invention to be carried by a person skilled in the art and may require the applicant to indicate the best mode for carrying out the invention known to the inventor at the filing date or where priority is claimed, at the priority date of the application". For all practical purposes, the same result is accomplished by the corresponding provision of Patent Co-operation Treaty(PCT), which includes describing the nucleotide sequences, deposition of micro-organism to supplement the written description. The provisions as mentioned above, also exist in the patent laws of almost all countries for the reason that when term of protection (patent) is over, the public should be able to take benefit of the invention. It is very difficult to describe the invention relating to biotechnology by written description as it involves the use of living material such as micro-organism. Budapest Treaty, provides facility to deposit the micro-organism in any of the International Depository Authority (IDA) recognized by WIPO to supplement the written description to avoid deposition of such micro-organism in each country where the applicant applies for grant of Patent.

Innovation of technology

One of the stated objectives of the TRIPS Agreement is to contribute to technological innovation. It is generally agreed that IP protection does so, although to what extent is unclear, and the role of IP systems in stimulating innovative activity should not be overstated. There are many other incentives–from prizes to subsidies and tax credits – that governments use to encourage the private sector to invest in research. For some innovators, the desire to gain a competitive advantage in a leading technological area is incentive enough to innovate. IPRs are, thus, only one factor contributing to technological innovation. Their impact is likely to change depending on other inter-relating factors, including the level of local research infrastructure, technical education, technological information and finance, the industry involved, the existence of coordinated government technology policy, and so on. Despite the complexity of these inter-relationships, a few observations can be made about the role of IP protection in encouraging innovation.

The Dissemination of Technology

According to the objectives of the TRIPS Agreement, the protection and enforcement of IPRs should also contribute to the transfer and dissemination of technology. IP systems can promote transfer and dissemination by providing an enabling environment for the licensing of IPRs, although empirical research on whether this is in fact, the case is inconclusive. In theory, guaranteed protection of IPRs could encourage risk-averse technology holders to set up joint ventures or wholly-owned subsidiaries in another country through which technology might be transferred. Alternatively, IPR protection can encourage companies to seek technology partners in other countries through the negotiation of licensing agreements. In particular, strong IPR protection could encourage small and medium-sized companies that lack the resources to set up subsidiaries off-shore and so, rely on licensing IPRs to introduce technology to new markets. Available statistics indicate that developed countries and transnational corporations currently dominate licensing activity and gain most in terms of royalty flows. The United Nations Conference on Trade and Development's (UNCTAD) World Investment Report notes that a high proportion of royalty payments and fee receipts accrue to only a few developed countries. A UNESCO report confirms this, indicating that only 10 developed countries capture 91% of all cross-border technology license fees. Further, technology flows are dominated by a limited number of transnational corporations. The UNCTAD Report noted that 70% of global royalties and license fees flowed between parent companies and their foreign affiliates. While the UNCTAD Report acknowledged small technology flows to developing countries, it concluded that the liberalization of regulatory policies on foreign technology agreements had not been sufficient to bridge the technology gap between developed and developing countries.

Benefits in Biological Science

Obligations under the TRIPS Agreement relevant to biotechnology

Article 27 of the TRIPS Agreement refers to "Patentable Subject Matter" and states that patents shall be available in all fields of technology. Article 27 includes some exceptions to this general rule, including Article 27(3) (b), which permits WTO Members, to exclude plants and animals from patentability; but requires them to:

- Recognize patents over micro-organisms;
- Recognize patents for microbiological processes for the production of plants and animals; and
- Recognize either patent or sui generis protection for plant varieties. With regard to the protection for plants varieties, Members have the option to:
- Provide *sui generis* protection for new plant varieties;
- Provide patent protection for new plant varieties; or
- Provide both *sui generis* and patent protection for new plant varieties.

With regard to the protection of innovations generally, Article 27(2) allows Members to exclude inventions from patentability to protect order public or morality, including to:

- Protect human life or health;
- Protect animal life or health;
- Protect plant life or health; or
- Avoid serious prejudice to the environment.

Such exclusions must not be made merely on the basis that the exploitation of the invention is prohibited under national law. In relation to the protection of plant varieties, what constitutes *sui generis* protection for new plant varieties is left undefined by the Agreement? One option is for countries to implement the International Convention for the Protection of New Varieties of Plants (UPOV). However, this form of protection has been criticized for focusing too much on the rights of plant breeders, and too little on the rights of those using the seeds–farmers. Members are encouraged to read Articles 27(2) and 27(3)(b) broadly, and to develop national systems that promote not only formal innovation by plant breeders, but also local breeding and the rights of farmers, and protect fundamental human rights, including the rights to food and health. Section two continues with a summary of the on-going debate concerning the relationship between the TRIPS Agreement and the Convention on Biological Diversity (CBD). The CBD sets as its objectives the conservation of biological diversity, the promotion of sustainable use of its components, and the fair and equitable sharing of the benefits arising out of the use of genetic resources. Implementation of these objectives requires the protection and use of knowledge relevant to the conservation and sustainable use of biological diversity, including knowledge about genetic material, knowledge in technology, there are still significant gaps in knowledge concerning the link between the award of private rights over intellectual property and the conservation and sustainable use of biodiversity. For example, the TRIPS Agreement obliges Members to introduce protection of new plant varieties, which are essentially property rights awarded to promote the breeding of new varieties. The introduction of new high yielding or easily cultivated species can lead to overplanting of one variety and the loss of others. To the extent that IPRs can be linked to the introduction of monocultures.

The requirement to protect new plant varieties could also lead to the establishment of IP systems that promote the innovation of genetically modified organisms (GMOs). As reflected in the negotiation by the international community of the Cartagena Protocol on Biosafety in January 2000, GMOs may pose a risk to the environment. In order to avoid future biodiversity loss associated with the introduction of untested GMOs, policy-makers, when considering whether and how to establish protection for plant varieties, should ensure that adequate regulatory systems exist to test and ensure the safety of any new GMOs.

Finally, the TRIPS Agreement could have an impact on the operation of the CBD as a result of decisions under the WTO's dispute settlement mechanism. While the CBD is legally binding, it contains no well-developed enforcement mechanism to address non-compliance by parties with their commitments. In contrast, TRIPS disputes are subject to the binding decision of the WTO's dispute settlement mechanism. In the case arising out of the implementation of the TRIPS Agreement, a WTO panel may give greater weight to the provisions of the TRIPS Agreement than to those of the CBD. To ensure such disputes are resolved consistently with the CBD's objectives, CBD experts should be included on dispute panels, and involved as independent experts, so that the WTO dispute settlement mechanism is adequately informed of all relevant evidence and jurisprudence.

Geographical indications are similar in function to trademarks, except that they identify a product with a particular territory, whereas trademarks identify a product with a company or brand. Members are required to permit legal action enabling traders to prevent (a) the designation or presentation of a good (such as a trademark) that suggests, in a manner that misleads the public, that the good in question originates in a geographical area other than the true place of origin; and (b) any use which constitutes unfair competition (Article 22). Article 23 deals solely with wines and spirits, which is indicative of the influence of the major wine and spirit-exporting countries in negotiating TRIPs. The geographic indications provisions are to be reviewed periodically by the Council for TRIPs (Article 24).

Recombinant Drugs, Vaccines and Diagnostics that are off patents or soon to be of patents.

Biogenetics' refers to therapeutic products based on genetically engineered or recombinant technologies that are already on the market at least in some industrialized countries. The first therapeutic protein produced through rDNA technology to be in the market, was Genentech's human insulin, introduced in 1982. The total amount of recombinant therapeutics molecules approved throughout the world is now around 30. In 2000, nearly 86 per cent of the 77 biotechnology medicines approved by the FDA (Food and Drug Administration of the U.S.A.) constitute recombinant human proteins. The approved products can be categorized into blood factors, hormones, growth factors, interferon, interleukins, vaccines, and other products. The estimated worldwide sales of recombinant products were US$ 1.4

billion in 1990 and US$ 6.6 billion in 2000. Over the next five years, more than $10 billion worth of products will come off patent. Many treatments for diseases like Diabetes, Gaucher's Disease, Hepatitis B&C, Sclerosis, Growth Hormone deficiency relying on biotechnology will face patent expiration between 2001 & 2005. In India, the market of approved recombinant therapeutics in 2001 was estimated to be about US$ 109 million, which represents 3.2% of the total Indian pharmaceutical market, and 1.6% of the world market for recombinant therapeutics.

Members should ensure that IP laws are clear in their rejection of overly broad claims that could lead to the grant of broad patents and excessive monopolies over areas of research. Members should review patenting requirements and elaborate standards with a view to promoting rather than stifling innovation. Article 30 of the TRIPS Agreement ensures that Members may provide limited exceptions to the exclusive rights conferred by a patent, provided that such exceptions do not unreasonably conflict with normal exploitation of the patent and do not unreasonably prejudice the legitimate interests of the patent owner. It is generally agreed that this includes the provision of research exemptions that allow third parties to use information disclosed in patent documents for research purposes without having to seek authorization of the patent owner. Policy-makers should ensure provision in national patent laws that implement research purposes. Developing countries should also make use of the technical assistance programs under Article 67 of the TRIPS Agreement to introduce state of the art information technology that will help researchers access patent information quickly and cheaply.

The TRIPS Agreement makes it mandatory to provide patent protection to micro-organisms and non-biological and microbiological production of plants and animals. Plant varieties are to be protected either by patents or by a *sui generis* system *or* a combination of both. This makes it difficult for the developing countries to exclude inventions within this category altogether. Hence, the strategy should be how to limit the scope of these provisions. As far as the patent protection of micro-organism is concerned, TRIPS does not provide a definition of micro-organism. The national rule-makers should define micro-organism in such a way as to include the following: bacteria, virus, fungus and algae. Another important way to limit the scope of patent protection to biological materials is to make a difference between the concept of invention and discovery. Since only inventions are qualified

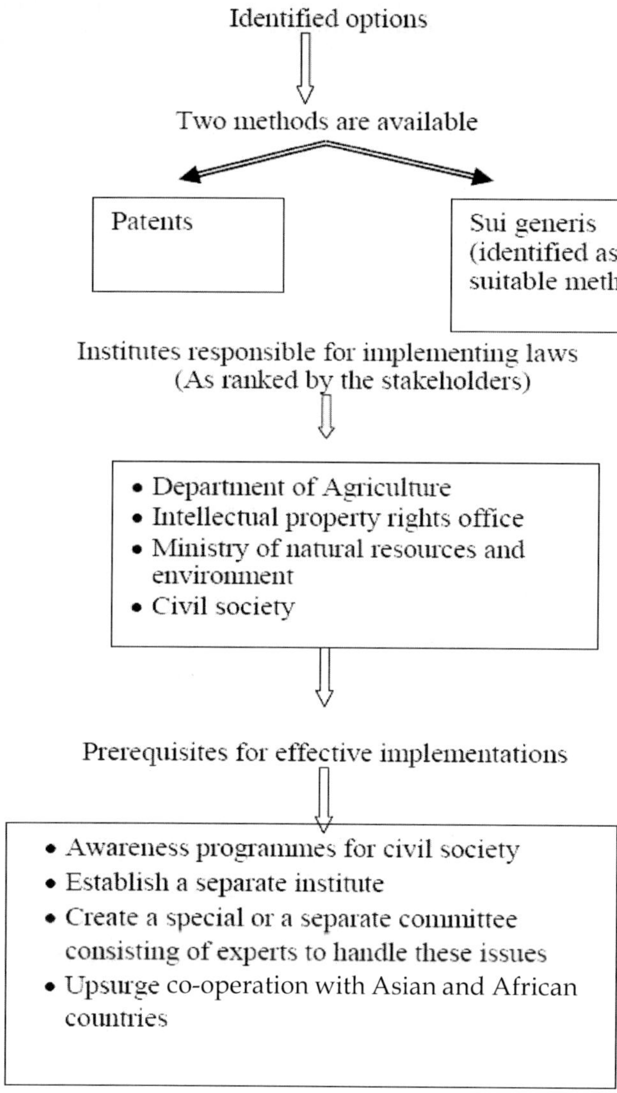

for patenting, naturally found micro-organisms, DNA structure, genes, blood cells, etc., can be excluded from patent protection. Developing countries can also exclude certain inventions in biotechnology by relying on the exclusion provision available under the TRIPS Agreement which permits the state parties to exclude certain inventions which are injurious to health and environment of human and animals. Using this exception, a member state can exclude terminator type technologies from patent protection.

Developing countries can use the *sui generis* option for the protection of plant varieties. The Agreement is silent about the content of the *sui generis* system. Hence, the developing countries can adopt a *sui generis* system which is suitable to their socio-economic conditions. There is no compulsion to adopt an UPOV model system. If the countries are going for the UPOV model, the successive amendments made the UPOV system are very stringent, especially after 1991 amendment. Therefore, the desirable model is UPOV 1978, which provides for breeders exception and also does not affect the farmer's rights. However, it is better to recognize the farmer's right explicitly. An explicit recognition would not give room for confusion in this matter. While doing so, care should be taken in outlining the rights of the farmers. It should not result in the curtailment of any rights enjoyed hitherto by the farmers. Further, the protection should be limited to certain varieties and avoid food grains from protection.

Animal Patents

In 1987, in a case involving polyploid oysters, the United States Patent and Trademark Office (USPTO) ruled that animals could be patented. In spite of the potential of transgenic animals as useful 'factories' for therapeutic proteins at a future date, current uses of such patented animals are as experimental models for screening assays, for example the Oncomouse (used to screen compounds for anti-cancer activity). In 1988, the Harvard Oncomouse was granted a Patent by USPTO. The European Patent Office, after much debate, granted the Patent in 1992. But it is incorrect to conclude that there is universal agreement on the patentability of the Oncomouse; for instance, Denmark has said 'No' to animal patents. The debate centers on ethical and moral issues, as well as animal rights.

Implications of the TRIPs Agreement on Biotechnology

Like in other disciplines, for biotechnology, protection of intellectual property provides encouragement for innovations involving genetic engineering, in addition to according incentives for investments, which may lead to new products and processes. The general pre-requisites for patentability, namely novelty, inventiveness and industrial applicability (or utility) apply to biotechnology inventions as well. As a rule, though this may appear paradoxical, new biological material is patentable, if obtained through non-biological processes. Non-biological processes are defined as those where the *hand of man* had a part to play. Article 27.3 (b) of TRIPs gives members the freedom to exclude plants, animals and "essentially biological processes" from patentability. However, it also states that micro-organisms and non-biological and micro-biological processes have to be patentable.

The wording is deliberately ambiguous, which gives countries some freedom to interpret this in their national legislation as they deem fit.

References

1. Simon Walker, 2001. The TRIPS agreement, sustainable development and the public interest-Discussion paper. IUCN Environmental Policy and Law Paper No. 41.
2. Manickavasagam V. 2007. Intellectual property rights and impact of TRIPS agreement with respect to Indian patent law. Project Report submitted to Planning Commission of India (SER Division), New Delhi, India.
3. Dilani Hirimuthugodage, 2011. Trade Related aspects of Intellectual Property Rights (TRIPS) and agriculture sector in Sri Lanka, Asia-Pacific Research and Training Network on Trade Working Paper Series, No. 92.

❑❑❑

Chapter 7

Protection of Traditional Knowledge within Modern IPR Era

Mukund A Kataktalware, Letha Devi G and K P Ramesha

Traditional Knowledge (TK) refers to knowledge systems, creations and innovations belonging to a particular community or people or its territory which is the outcome of intellectual activity and insight in a traditional context that is transmitted between generations and is an adaptation to a changing environment. World Intellectual Property Organization (WIPO, 2012) summarizes TK as a living body of knowledge that is developed, sustained, and passed on from generation to generation within a community, often forming part of its cultural or spiritual identity. In broad sense TK includes the intellectual and intangible cultural heritage, practices, and knowledge systems of indigenous and local communities. In a narrower sense, it refers to knowledge as such, in particular the "content or substance of knowledge resulting from intellectual activity in a traditional context", which includes the technical know how, innovations, practices, skills, and learning, and knowledge symbolizing traditional way of life of indigenous and local communities, or contained in codified knowledge systems passed from generation to generation. It is not restricted to any specific technical field, and may include agricultural, environmental, health care of humans as well as animals, and genetic resources. TK is a part and parcel of the cultural identities of indigenous communities. It not only plays a significant role in livelihood of millions of people living in developing and underdeveloped countries but also has significance for their future well-being and sustainable development. In spite of spread of modern medicine worldwide, still sizable population in developing world relies on traditional system of medicine which includes Ayurveda, Unani, Siddha, Chinese, Tibetian, Homeopathy, Yoga, Meditation, Acupuncture, etc. Similarly, ethno-veterinary practices are preferred by several indigenous communities for health care of their livestock.

The wider application of TK indicates that it is an inseparable part of the indigenous community. Belatedly due to the exploitation of such invaluable knowledge base at the cost of owner's rights concerns are being expressed to protect TK using intellectual property rights (IPRs). The traditional communities, who developed these knowledge bases, are often sidelined in the process of protecting IPRs. Big multinational companies tend to exploit undocumented TK by patenting the same for commercial gains. There is an urgent need to ensure that the wealth of TK is utilized for the benefit of local communities who owns it and is not misappropriated by the vested interests.

Significance of Ethno-Veterinary Knowledge

Traditional indigenous knowledge focusing on ethno-veterinary medicine has existed alongside human evolutionary history. In India, the earliest information on the art of caring for animals was provided by the sacred texts of the Vedic religion (Tiwari and Pande, 2010). The Atharvaveda (IV,9,2) hails the benefits of protective ointment for human beings, cows and horses (Mazars, 1998). The era of treating ethno-veterinary medicine and any other ethno-knowledge system with suspicion and labelling it as myth, superstition and witchcraft, is long gone (Wanzala *et al.*, 2005). On the contrary the role of ethno-veterinary medicine in livestock development is now acknowledged worlwide. The livestock sector in India is one of the largest in the world (56.7% of world's buffaloes, 12.5% cattle, 20.4% small ruminants, 2.4% camel, 1.4% equine, 1.5% pigs and 3.1% poultry). The prevention and control of livestock diseases and improving productivity of this genetic resource is a big challenge. The application of scientifically validated ethno-veterinary practices may help in addressing this problem.

Forms of TK: Traditional Knowledge systems are complex, varied and dynamic, and they find expression in different forms. The classification of the different forms in which TK may be found, is significant for an analysis of the relationship between intellectual property (IP) and TK. As per the WIPO (2010), TK may be categorized as follows:

Unfixed TK and Fixed TK: Some TK is "fixed", meaning that it is recorded in "a sufficiently stable" material or tangible form. Fixed TK can be either verbal" (such as TK recorded in a song, a book or film; and traditional norms and recipes) or "non-verbal" (such as TK embodied in traditional architecture or rock art). "Verbal" means "finding expression in words only, without being manifested in action" Fixed TK could be either written (such as a written document) or non-written (such as a sound recording). "Written" means "composed, recorded, preserved, or mentioned in writing". Some TK is recorded and preserved in writing, while some TK is fixed in video, audio and in some other kinds of non-written forms. "Unfixed" TK refers to TK which is not recorded in tangible form. Unfixed TK could be found orally (such as in the form of "oral history") and in other non-written forms (such as music, performance and actions). Unfixed TK can be either verbal or non-verbal. Healing skills and techniques, which might be transmitted orally, could be unfixed "verbal" TK, while TK embodied in traditional performances or music (without songs or other verbal forms) would be unfixed "non-verbal" TK.

Documented TK and Non-Documented TK: Fixed TK can be either documented or non-documented. To document is to record, for purposes of furnishing evidence. Documented TK can take many forms, such as text, video, audio, etc., or a combination thereof. Documented TK can be either verbal or non-verbal, and can be either written or non-written. However, most TK remains largely non-documented in this sense of the term.

Codified TK and Non-Codified TK: TK can be either codified or non-codified. Codified TK is that which is in some systematic and structured form, in which the knowledge is ordered, organized, classified and categorized in some manner. Codified TK may be imbued with some "authority" or legitimacy.

Disclosed TK and Undisclosed TK: TK can be disclosed through use, orally or through its documentation and the dissemination of that documentation.

TK Directly Controlled by Indigenous and Local Communities and TK No Longer in the Control of Indigenous and Local Communities: This distinction might be said to be related to the concepts of "disclosed" and "undisclosed" TK.

TK Held by Indigenous and Local Communities: The notion "holder" and "held by" are used in the CBD and its Nagoya Protocol on Access to Genetic Resources and the Fair and Equitable Sharing of Benefits Arising from their Utilization.

Sacred TK and Secular TK: "Sacred" refers to "any expression of TK that symbolizes or pertains to religious and spiritual beliefs, practices or customs. It is used as the opposite of profane or secular, the extreme forms of which are commercially exploited forms of TK." Sacred TK refers to the TK which includes religious and spiritual elements, such as totems, special ceremonies, sacred objects, sacred knowledge, prayers, chants, and performances and sacred symbols, and also refers to sacred TK associated with sacred species of plants, animals, micro-organisms, minerals, and refers to sacred sites. Whether TK is sacred or not depends on whether it has sacred significance to the relevant community. Much sacred TK is by definition not commercialized, but some sacred objects and sites are being commercialized by religious, faith-based, and spiritual communities themselves, or by outsiders to these, and for different purposes.

TK "as such" and TK-based Innovations and Creations: TK "as such" refers to knowledge systems, creations, and innovations which have generally been transmitted from generation to generation, are generally regarded as pertaining to a particular people or its territory; and, are constantly evolving in response to a changing environment. While it is often thought that tradition is only about imitation and reproduction, it is also about innovation and creation within the traditional framework. Tradition is not immutable. TK itself is the result of intellectual creativity and creations within a "traditional context." "TK-based innovations and creations" refer to innovations and creations based on TK "as such," developed and innovated beyond a *"traditional context"*. Such innovations and creations may be done by indigenous and local communities, or certain members thereof, but they can also be done by third parties beyond such communities. TK-based innovations and creations do not arise within a "traditional context", in other words, they are not: "(i) generated, preserved and transmitted in a traditional and inter-generational context; (ii) distinctively associated with a traditional or indigenous community or people which preserves and transmits it between generations; and (iii) integral to the cultural identity of an indigenous or traditional community or people which is recognized as holding the knowledge through a form of custodianship, guardianship, collective ownership or cultural responsibility."

Indigenous Knowledge and Traditional Knowledge: As indicated in the WIPO Report on the Fact-Finding Missions on Intellectual Property and Traditional Knowledge (1998-1999), cited above, "indigenous knowledge" may be used to describe knowledge held and used by communities, people and nations that are "indigenous". In this sense, "indigenous knowledge" would be the TK of indigenous people. Indigenous knowledge is, therefore, a part of the TK category, but TK is not necessarily indigenous. In Committee documents, it has been proposed that "traditional" means "(i) generated, preserved and transmitted in a traditional and inter-generational context; (ii) distinctively associated with a traditional or indigenous community or people which preserves and transmits it between generations; and (iii) integral to the cultural identity of an indigenous or traditional community or people which is recognized as holding the knowledge through a form of custodianship, guardianship, collective ownership or cultural responsibility".

Individual TK and Collective TK: TK is, in general, developed collectively and/or regarded as belonged collectively to an indigenous or local community or to groups of individuals within such a community. National legislations in some countries have provisions recognizing collective ownership or custodians of TK, especially for indigenous and local communities and specifically regarding biodiversity or genetic resources. Nonetheless, a particular individual member of a community, such as a certain traditional healer or individual farmer, might hold specific knowledge. This is recognized in some national and regional laws.

Commercialized TK and Non-Commercialized TK: Some TK has already been commercialized. To "commercialize" may be defined as to "exchange goods and services from the point of production to the point of consumption." Commercialization is done within a commercial or monetary system. TK might be commercialized as part of local industrial traditions or community enterprises, or as part of a business agreement with a third party. However, much TK has not been commercialized.

Importance of Protecting Traditional Knowledge

Traditional knowledge is vital for existence of indigenous and local communities which depend on TK for their health, livelihoods, and general well-being. Thus, a TK regime that facilitates and promotes the conservation and continued use of TK relating to health and food production could potentially improve the standard of life of millions of people. TK is used as an input into modern industries, such as pharmaceuticals, botanical medicines, cosmetics, agriculture and biological pesticides, etc. Some traditional medicines are used as inputs in biomedical research, suggesting that they may constitute a source of income not just as drugs in themselves but as the sources of chemical substances forming the basis of new pharmaceuticals (Dutfield, 2006). Even though indigenous communities play an important role in the unearthing, development, and preservation of various medicinal plants, health-giving herbal formulations, agricultural and forest products, they are largely deprived of any benefits accruing to their contribution. This injustice meted out to traditional communities needs to be addressed so that those communities and countries to which they belonged to get benefitted from the value addition which multinational companies reap by investing drug discovery, development and marketing. Thus, protecting TK has the inherent potential to improve the performance of many developing-country economies by enabling greater commercial use of their natural wealth. Numerous academic studies on indigenous communities indicate that the protection of TK can help in conservation of environment. Further, protection of TK may help in prevention of biopiracy, i.e. unauthorised exploitation of biological resources without adequate compensation to the owners of TK. The well known examples of biopiracy of traditional knowledge include futile attempts to patent certain compounds/traits/qualities of turmeric, neem, Basmati rice which were foiled by the aggressive approach of the Indian authorities.

International Developments

WTO Ministerial Conference, Doha (2001): A key issue raised during this conference was the TRIPS Agreement requires to be amended in order to provide that members shall require an applicant for a patent relating to biological material or traditional knowledge shall disclose the source and country of origin of the biological resource and of the traditional knowledge used in the invention, evidence of prior informed consent and evidence of fair and equitable benfit sharing, as a precondition to acquiring patent rights.

Conference of Parties to Convention of Biological Diversity Member, Hague (2002): The Convention on Biological Diversity (CBD), signed at the United Nations Conference on Environment and Development (UNCED) in 1992, was the first international environmental convention to develop measures for the use and protection of traditional knowledge, related to the conservation and sustainable use of biodiversity. The requests made by the Conference of Parties (COP6) in The Hague conference in 2002 were to address the issues of *sui generis* systems for the protection of traditional knowledge based on Article 8(j) and related provisions of the CBD. The parties to the Convention set a 2010 target to negotiate an international legally binding regime on access and benefit sharing (ABS) at the Eighth meeting (COP8), 20–31 March, 2006 in Curitiba, Brazil. This target was met in October 2010 in Nagoya, Japan, by conclusion of the Nagoya Protocol to the CBD. A major accomplishment at COP 10 was the adoption of the Nagoya Protocol on Access to Genetic Resources and the Fair and Equitable Sharing of Benefits Arising from the Utilization of Genetic Resources of the Convention on Biological Diversity (Nagoya Protocol) following six years of intense negotiations. Indigenous and local communities participated, including those on government delegations, throughout the negotiations. The Nagoya Protocol is the first international instrument of particular relevance to indigenous and local communities negotiated since the adoption of the UN Declaration on the rights of indigenous people (September 2007). As such, it is a significant step in mainstreaming indigenous rights as a cross-cutting issue in international negotiations. The purpose of the Protocol is to effectively implement one of the three core objectives of the Convention: the fair and equitable sharing of benefits arising from the utilization of genetic resources. It builds on the access and benefit-sharing provisions of the Convention.

Global Biodiversity Forum, Cancun, 2003: At the 18th conference of the Global Biodiversity Forum (GBF 18) held at Cancun, Mexico in 2003, biodiversity and sustainable livelihood issues related to international trade were discussed. The GBF emphasized that economic growth cannot take place at the cost of environment.

Tools for Protecting TK

The IP system can be approached from two different directions to ensure protection of TK. These two approaches are generally referred to as "**defensive**" and "**positive**" protection and can be undertaken together in a complementary way. Defensive protection of TK prohibits others from obtaining IPR of holder of TK. Some countries like India have developed a searchable TK database, which may be used as an evidence of prior art to defeat a claim to a patent based on such TK. Under the positive protection of TK, the IP systems is designed to enable TK holders, to acquire and assert IP rights in their TKs, with the resulting possibility of preventing others for using the TK without permission. In some countries, the positive protection of TK can be availed under *sui generis* legislation. Thus, positive protection is the granting of rights that empower traditional communities to promote their TK, manage their uses by third parties and gain from their commercial exploitation.

Indian Efforts towards TK Protection

The Traditional Knowledge Digital Library (TKDL) was set up by the Government of India in 2001, as repository of 1200 formulations of various systems of Indian medicine and 1500 Yoga postures (*asanas*) and translated the subject matter into five languages *viz.*, English, French, German, Japanese and Spanish. Further, India has signed various agreements with United Kingdom Intellectual Property Office (UKIPO), the European Patent Office (EPO), and the United States Patent and Trademark Office (USPTO) to prevent the grant of invalid patents by giving patent examiners at International Patent Offices access to the TKDL database for patent search and examination. India made a strong representation in 2004 for inclusion of Indian Journal of Traditional Knowledge and Medicinal and Aromatic Plant Abstracts for including in Non-Patent Literature (NPL) list of periodicals, which was acceded to and considered as a major breakthrough since this was for the first time two periodicals from developing countries were included in the NPL list.

Conclusion

Traditional knowledge is vital for existence of indigenous and local communities which depend on TK for their health, livelihoods, and general well-being. Thus, a TK regime that facilitates and promotes the conservation and continued use of TK relating to health and food production could potentially improve the standard of life of millions of people. Thus, protecting TK by defensive as well as positive means has the inherent potential to improve the performance of many developing-country economies by enabling greater commercial use of their natural wealth. The enabling regulatory framework may facilitate judicious use of traditional knowledge for the benefit of mankind in general and traditional communities in particular.

Traditional Knowledge Laws in India

Title	Patents Act, 1970 as amended by Patents (Amendments) Act, 2005	Biological Diversity Act, 2002
Subject Matter	Traditional Knowledge	Traditional Knowledge
Issue(s)	Scope of Protection	Beneficiaries or Rightholders; Exceptions and Limitations
Type(s) of Legislation	IP Law	Other
Scope of Protection	Section 3 - What are not inventions The following are not inventions within the meaning of this Act,- (p) an invention which, in effect, is traditional knowledge or which is an aggregation or duplication of known properties of traditionally known component or components. Section 25 (2) At any time after the grant of patent but before the expiry of a period of one year from the date of publication of grant of a patent, any person interested may give notice of opposition to the Controller in the prescribed manner on any of the following grounds, namely: – (k) that the invention so far as claimed in any claim of the complete specification was anticipated having regard to the knowledge, oral or otherwise, available within any local or indigenous community in India or elsewhere but on no other ground. (3) (a) Where any such notice of opposition is duly given under sub-section (2), the Controller shall notify the patentee. (b) On receipt of such notice of opposition, the Controller shall, by order in writing, constitute a Board to be known as the Opposition Board consisting of such officers as he may determine and refer such notice of opposition along with the documents to that Board for examination and submission of its recommendations to the Controller. (c) Every Opposition Board constituted under clause (b) shall conduct the: examination in accordance with such procedure as may be prescribed. (4) On receipt of the recommendation of the Opposition Board and after giving the patentee and the opponent an opportunity of being heard, the Controller shall order either to maintain or to amend or to revoke the patent.	Beneficiaries or Rightholders Section 2 (a) "benefit claimers" means the conservers of biological resources, their byproducts, creators and holders of knowledge and information relating to the use of such biological resources, innovations and practices associated with such use and application. Exceptions and Limitations Section 5 (1) Sections 3 and 4 not to apply to certain collaborative research projects. The provisions of sections 3 and 4 shall not apply to collaborative research projects involving transfer or exchange of biological resources or information relating thereto between institutions, including Government sponsored institutions of India, and such institutions in other countries, if such collaborative research projects satisfy the conditions specified in sub-section (3). (2) All collaborative research projects, other than those referred to in sub-section (1) which are based on agreements concluded before the commencement of this Act and in force shall, to the extent the provisions of agreement are inconsistent with the provisions of this Act or any guidelines issued under clause (a) of sub-section (3), be void. (3) For the purposes of sub-section (1), collaborative research projects shall- (a) conform to the policy guidelines issued by the Central Government in this behalf; (b) be approved by the Central Government.

Adapted from http://www.wipo.int/tk/en/databases/tklaws/articles DOA 08/08/2014

References

Dutfield, G. 2006. Protecting traditional knowledge: Pathways to the future. Published by International Centre for Trade and Sustainable Development (ICTSD) International Environment House 2 7 Chemin de Balexert, 1219 Geneva, Switzerland.

Mazars, G. 1998. Veterinary medicines in ancient and medieval India. *Studies His. Med. Sci.*, 16(1-2): 27-36.

Tiwari, L and Pande, P.C. 2010. Ethno-veterinary medicines in Indian perspective: Reference to Uttarakhand, Himalaya. *Indian Journal of Traditional Knowledge*, 9(3): 611-617.

Wanzala, W., Zessin, K.H., Kyule, N.M., Baumann, M.P.O., Mathias, E. and Hassanali, A. 2005. Ethno-veterinary medicine: a critical review of its evolution, perception, understanding and the way forward. *Livestock Research for Rural Development,* 17 (11):

World Intellectual Property Organization. 2012. Intellectual property and genetic resources, traditional knowledge and traditional cultural expressions: An overview. Published by World Intellectual Property Organization, 34, chemin des Colombettes, P.O. Box 18, CH-1211 Geneva 20, Switzerland.

World Intellectual Property Organization. 2010. Inter-governmental Committee on Intellectual Property and Genetic Resources, Traditional Knowledge and Folklore-Seventeenth Session. Published by World Intellectual Property Organization, 34, chemin des Colombettes, P.O. Box 18, CH-1211 Geneva 20, Switzerland.

□□□

Chapter 8

Indian Legislations for the Protection of Various Types of Intellectual Properties

Mukund A Kataktalware, Letha Devi G and K P Ramesha

Intellectual Property (IP) is a kind of property which confers rights over intangible creations of human mind. Intellectual Property Rights (IPRs) are legal rights granted to creators and owners of IP, which encompasses works of human intellectual creativity. The Convention Establishing the World Intellectual Property Organization (WIPO) defines IP as the rights relating to literary, artistic and scientific works; performances of performing artists, phonograms, and broadcasts; inventions in all fields of human endeavour; scientific discoveries; industrial designs; trademarks, service marks, and commercial names and designations; protection against unfair competition; and all other rights resulting from intellectual activity in the industrial, scientific, literary or artistic fields. IPRs usually give the creator/owners of IP the right to prohibit others from unauthorised exploitation of IP for a defined period of time. IP laws provide the enabling environment that promotes innovation and creativity in all walks of life. The enforcement of IPRs significantly contributes to technological advancement, competitiveness in commerce and overall socio-economic development of the country.

Development of Intellectual Property Law in India

The first Indian Patent Law was enacted in 1856, followed by Indian Patents and Designs Act in 1911 and Indian Copyright Act in 1914. In 1948, the Indian Government appointed the first committee to review the prevailing Patents and Designs legislation. In 1957, Government appointed Justice Rajagopal Ayyangar Committee to revise the Patent Law. Rajagopal Ayyangar Committee submitted its report on 1959, the report tried to balance the constitutional guarantee of economic and social justice enshrined in the preamble of the constitution. Another expert committee headed by Bakshi Tek Chand also reviewed the existing situation. The Patent Bill, following the reports of these committees was debated for a decade leading to enactment of Indian Patent Act, 1970. Since then the Government of India enacted several landmark legislations which are presented in Table 1.

Table1. Different IP Laws/ Regulations enacted by the Government of India

Type of Laws/ Regulations	Name of Laws/ Regulations (Date of current version)
Main IP Laws	• The Patents Act, 1970 (as amended up to Patents (Amendment) Act, 2005) (2013) • The Copyright (Amendment) Act, 2012 (2012) • The Trade Marks (Amendment) Act, 2010 (2010) • Patents (Amendment) Act, 2005 (Act No. 15 of 2005) (2005) • Patents (Amendment) Act, 2002 (2002) • Protection of Plant Varieties and Farmers' Rights Act, 2001 (2001) • The Semi-conductor Integrated Circuits Layout-Design Act, 2000 (2000) • The Designs Act, 2000 (2000) • Copyright (Amendment) Act, 1999 (Act No. 49 of 1999) (1999) • The Geographical Indications of Goods (Registration and Protection) Act, 1999 (1999) • Patents (Amendment) Act, 1999 (1999) • Copyright Act, 1957 (1999) • Copyright (Amendment) Act, 1994 (1994)
IP-related Laws	• Biological Diversity Act, 2002 (2002) • Drugs and Cosmetics Act, 1940 (1995) • The Seeds (Amendment) Act, 1972 (1972) • The Seeds Act 1966 (1966)
Implementing Rules/ Regulations	• Patents (Amendment) Rules, 2014 (2014) • The Semi-conductor Integrated Circuits Layout-Designs Act, 2000 (Notification) (2014) • Gazette Notification of Trade Marks (Amendment) Rules, 2013 and Trade Marks (Amendment) Act, 2010 (2013) • Patents (Amendment) Rules, 2013 (2013) • Trade Marks (Amendment) Rules, 2013 (2013) • The Copyright Rules, 2013 (2013) • The Patents Rules, 2003 (2013) • Patents (Amendment) Rules, 2012 (2012) • Trade Marks (Amendment) Rules, 2010 (2010) • Trade Marks (Second Amendment) Rules, 2010 (2010) • Protection of Plant Varieties and Farmers Rights (Criteria for DUS for Registration) • Regulations, 2009 (2009) • Protection of Plant Varieties and Farmers' Rights (Third Amendment) Rules, 2009 (2009) • Protection of Plant Varieties and Farmers' Rights (Second Amendment) Rules, 2009 (2009)

- Designs (Amendment) Rules, 2008 (2008)
- Circular on Implementing the Intellectual Property Rights (Imported Goods) Enforcement Rules, 2007 (2007)
- Intellectual Property Rights (Imported Goods) Enforcement Rules, 2007 (2007)
- Protection of Plant Varieties and Farmers' Rights Regulations, 2006 (2006)
- Patent (Amendment) Rules, 2006 (2006)
- Patents (Amendment) Rules, 2005 (2004)
- Biological Diversity Rules, 2004 (2004)
- Protection of Plant Varieties and Farmer's Rights Rules, 2003 (2003)
- The Geographical Indications of Goods (Registration and Protection) Rules, 2002 (2002)
- Trademarks Rules, 2002 (2002)
- Semiconductor Integrated Circuits Layout-Design Rules, 2001 (2001)
- The Designs Rules, 2001 (2001)
- The International Copyright Order, 1999 (1999)
- The Cinematograph Act, 1952 (1984)

Copyrights and Related Rights

Copyrights and Related Rights are the set of exclusive rights granted to the author or creator in respect of original literary, musical, artistic or audio-visual works. The Copyright Act of 1957, The Copyright Rules, 1958 and the International Copyright Order, 1999 governs the copyright protection in India. The Act has been amended in 1983, 1984, 1992, 1994, 1999 and 2012. The 'Related rights or Neighbouring rights' are the rights of performers (e.g., actors, singers and musicians), producers of phonograms (sound recordings) and broadcasting organizations. Indian Copyright Act provides separate and exclusive copyright protection to original literary work, original dramatic work, original musical work, original artistic work, cinematograph films, sound recording, and computer programme.

The Department of Higher Education, Ministry of Human Resource Development, Government of India administers the Act. A Copyright Board established under the Act is entrusted with the task of adjudication of disputes pertaining to copyright registration, assignment of copyright, grant of licences in respect of works withheld from public, unpublished Indian works, production and publication of translations and works for certain specified purposes. The Act set up a Copyright Office for the registration of Copyrights, which functions under the control of Registrar of Copyrights.

Procedure for registration : Application for Copyright Registration is to be made on Form IV as prescribed in the schedule I to the Rules.

Separate applications should be made for registration of each work. Each application should be accompanied by the requisite fee prescribed in the schedule II to the Rules. The applications should be signed by the applicant or the advocate in whose favour a *Vakalatnama* or Power of Attorney has been executed. The Power of Attorney signed by the party and accepted by the advocate should also be enclosed.

Duration of protection : In the case of literary, dramatic, musical or artistic works, the sixty years period is counted from the year following the death of the author. In the case of cinematograph films, sound recordings, photographs, posthumous publications, anonymous and pseudonymous publications, works of government and public undertakings and works of international organisations, the 60-years period is counted from the date of publication. In case of Broadcast reproduction right-25 years from the beginning of the calendar year next following the year in which the broadcast is made. In case of Performers right-25 years from the beginning of the calendar year next following the year in which the performance is made.

Copyright infringement : Copyright in a work shall be deemed to be infringed, if any person, without a licence granted by the owner of the Copyright or the Registrar of Copyrights under this Act or in contravention of the conditions of a licence so granted or of any conditions imposed by a competent authority under this Act: (i) does anything, the exclusive right to do which is by this Act conferred upon the owner of the copyright; or (ii) permits for profit any place to be used for the communication of the work to the public, where such communication constitutes an infringement of the copyright in the work, unless he was not aware and had no reasonable ground for believing that such communication to the public would be an infringement of copyright; or when any person: (i) makes for sale on hire, or sells or lets for hire, or by way of trade displays or offers for sale or hire; or (ii) distributes either for the purposes of trade or to such an extent as to affect prejudicially the owner of the copyright; or (iii) by way of trade exhibits in public; or (iv) imports into India, any infringing copies of the work. In Indian law, the minimum punishment for copyright infringement is imprisonment for six months with the minimum fine of Rs 50,000. In the second and subsequent conviction, the minimum punishment is one year imprisonment and fine of Rs. 1,00,000.

Patents

A patent is a legal monopoly granted by the Government to the owner of an invention for a limited period, in exchange of full disclosure of his invention. It consists of an exclusive right to produce the new invented article or produce an article according to the invented process for a pre-defined period. The Patent System in India is governed by the Patents Act, 1970 as amended by the Patents (Amendment) Act, 2002 and 2005 and the Patents Rules, 2003, as amended by the Patents (Amendment) Rules 2006 and 2014. After the amendments, product patent (instead of process patent) is being granted for food, pharmaceutical and chemical products. Also, along with post grant opposition to patents, pre-grant opposition is also permissible. Patent system in India is administered under the superintendence of the Controller General of Patents, Designs, Trademarks and Geographical Indications under the Department of Industrial Policy and Promotion, Ministry of Commerce and Industry. The Controller General directs and supervises the functioning of the Patent Office and the Patent Information System (PIS). The Head Office of Patents is at Kolkata with branches at Mumbai, Chennai and Delhi. The branches deal with the applications for patents originating within their respective territorial jurisdiction. The Patent Information System (PIS) at Nagpur has been functioning as patent information base for the users.

Patent application: Every application for a patent shall be for one invention only and shall be made in the prescribed form and filed in the patent office. Every application shall state that the applicant is in possession of the invention and shall name the owner claiming to be the true and first inventor; and where the person so claiming is not the applicant or one of the applicants, the application shall contain a declaration that the applicant believes the person so named to be the true and first inventor. Every such application shall be accompanied by a provisional or a complete specification. If a patent application is accompanied by a provisional specification, a complete specification shall be filed within twelve months from the date of filing of the application, and if the complete specification is not so filed, the application shall be deemed to be abandoned. At any time within four months from the date of advertisement of the acceptance of a complete specification under this Act, any person interested, may give notice to the Controller of opposition to the grant of the patent on certain permissible grounds.

Duration of protection: Term of every patent will be 20 years from the date of filing of patent application, irrespective of whether it is filed with provisional or complete specification. Date of patent is the date on which the application for patent is filed. The term of patent in case of International applications filed under the Patent Co-operation Treaty Designating India, will be 20 years from the International filing date accorded under the Patent Cooperation Treaty. A patent will have cease to effect on the expiration of the period prescribed for the payment of any renewal fee, if that fee is not paid within the prescribed period.

Infringement of patents: Infringement of a patent consists of the unauthorized making, importing, using, offering for sale or selling any patented invention within the India. The remedies against infringement of a patented invention include inter-locutory injunction, relief of damages, and accounts of profits.

Trademarks

Any distinctive symbols/signs that help the consumers of the product/service to distinguish between competing goods or services constitutes a trade and service marks. The trademarks is one of the important valuable forms of intellectual property rights and is commonly used by organizations, companies, businesses, institutions globally to enhance their image. In 1940, the then British Government of India passed the Trademarks Act for uniform and systematic registration of trademarks in India, which came into force on 1st June, 1942. This act was replaced by the Trade and Merchandise marks Act, 1958, which came into effect from 25th November, 1959. India, being the original member of the WTO, it was obligatory for India to amend the Trade and Merchandise marks Act, 1958 in conformity with the TRIPS. Therefore, the Government of India enacted The Trademark Act, 1999. This Act is in harmony with two major international treaties on the subject, namely The Paris Convention for Protection of Industrial Property and TRIPS Agreement to both of which India is a signatory.

The Trademark Act, 1999 and the Trade Marks Rules, 2002 govern the law relating to Trademarks in India. All the provisions of the Trademarks Act, 1999 and the Trademarks Rules came into effect on 15th September, 2003. According to the Trademarks Act, 1999, the term 'trademark' means "a mark capable of being represented graphically and which is capable of distinguishing the goods or services of one person from those of the others and may include shape of goods, their packaging and combination of colours".

Under the Act, the Controller-General of Patents, Designs and Trademarks under Department of Industrial Policy and Promotion, Ministry of Commerce and Industry is the 'Registrar of Trademarks'. The Controller General of Patents, Designs and Trademarks directs and supervises the functioning of the Trademarks Registry (TMR). The Registration of trademarks, is done in Mumbai, where the Trademarks Registry is situated. There are branch offices of the Registry at Kolkata, Delhi, Chennai and Ahmedabad with separate jurisdictions. The 'Trademarks Registry' administers the Trademarks Act, 1999 and the Rules thereunder. TMR acts as a resource and information centre and is a facilitator in matters relating to trademarks in the country. The main function of the Registry is to register trademarks which qualify for registration under the Act and Rules.

Procedure for registration: Applications for trademark registration are evaluated mainly with respect to the distinctiveness, possibility of deceptiveness and conflicting trademarks. The Registrar on consideration of the application and any evidence of use or distinctiveness decides whether the application should be accepted for registration or not, and if accepted, publishes the same in the official gazette i.e. Trademarks Journal. Within a defined period, any person can file an opposition, a copy of which is served to the applicants who is required to file a counter statement within two months failing which the application shall be treated as abandoned. Thereafter, the opponent leads evidence in support of his case by way of affidavit followed by the applicant's evidence also by way of affidavit in support of the application. After that, the opponent files evidence by way of rebuttal. On completion of evidence, the matter is set down for a hearing and the case is decided by a Hearing officer. The Registrar's decision is appealable to the Intellectual Property Appellate Board.

Duration of protection: The registration of trademark, after the commencement of this Act, shall be for a period of ten years, but may be renewed from time to time by paying prescribed fee.

Industrial Designs

An industrial design is the ornamental or aesthetic aspect of an article, which may include features of shape, configuration, pattern, composition of lines or colours whether in two dimensional or three dimensional or in both forms-by any industrial process or means, whether manual, mechanical or chemical, separate or combined, which in the finished article appeal to and are judged solely by the eye. The Designs Act, 2000 and Design Rules, 2001 and 2008 govern the design law in India. Under the Act, the Controller-General of Patents, Designs and Trademarks under Department of Industrial Policy and Promotion, Ministry of Commerce and Industry is the Controller of Designs. The Controller General of Patents, Designs & Trademarks directs and supervises the functioning of 'Industrial Designs Wing'. The registration of industrial designs under the Designs Act is done by the 'Industrial Designs Wing' of the Head Office of Patents located at Kolkata.

Registration: The Controller may, on the application of any person claiming to be the proprietor of any new or original design not previously published in any country and which is not contrary to public order or morality, register the design under the Act. Every application shall be in the prescribed form and shall be filed in the Patent Office in the prescribed manner and shall be accompanied by the prescribed fee. A design may be registered in not more than one class, and, in case of doubt as to the class in which a design ought to be registered, the Controller may decide the question. The Controller shall grant a certificate of registration to the proprietor of the design when registered. Also, there shall be kept at the Patent Office a book called the 'register of designs', wherein shall be entered the names and addresses of proprietors of registered designs, notifications of assignments and of transmissions of registered designs, and such other matter, as may be prescribed.

Duration: When a design is registered, the registered proprietor of the design shall, subject to the provisions of this Act, have copyright in the design during ten years from the date of registration. If, before the expiration of the said ten years, application for the extension of the period of copyright is made to the Controller in the prescribed manner, the Controller shall, on payment of the prescribed fee, extend the period of copyright for a second period of five years from the expiration of the original period of ten years.

Geographical Indications (GI)

GIs are signs or expressions used to indicate that a product *viz.*, agricultural goods, natural goods or manufactured goods originates in a country/region or specific place, where a given quality or reputation or other unique characteristics of the product are due to its geographical origin. The Geographical Indications (GI) of Goods (Registration and Protection) Act, 1999 came into force in September 2003. This law was enacted to comply with India's obligations under the TRIPs Agreement. GI can be registered by an association of persons or producers or by an organisation or an authority established under law for the time-being in force representing the interest of the producers under the Geographical Indications of Goods (Registration and Protection) Act, 1999. GI is registered for a period of 10 years and can be renewed from time to time. Since the Darjeeling tea became the first GI tagged product in India, in 2004-05, a total 215 products were registered as GI till March, 2014. Other examples of GI registered in India include Tirupati Laddu, Puneri Pagadi, Dharwad Pedha, Mysore sandalwood oil, Kullu shawl, etc.

Acknowledgement : The authors acknowledge the use of IPR literature available on internet in particular **www.business.gov.in/,** www.wipo.int **and** www.metastudio.org for compilation of this paper.

Conclusion

India, being a signatory of Agreement on Trade Related Aspects of Intellectual Property Rights, has evolved sophisticated administrative and legislative framework for protection of IPRs. These laws are expected to play a catalytic role in promoting innovation and creativity in all walks of life. The enforcement of IPRs would facilitate the technological advancement, competitiveness in commerce and overall socio-economic development of the country.

References

Adukia, R. 2012. Handbook of Intellectual Property Rights in India. Retrieved August 13, 2014 from

http://www.metastudio.org/Science%20and%20Ethics/file/readDoc/535a76367d9d331598f49e2d/34_Hb_on_IPR.pdf

Important web resource:

Intellectual Property Office, India: www.ipindia.nic.in

Patent office, India: www.patentoffice.nic.in

Copyright Office, India: http://copyright.gov.in

Automated Recordation & Targeting for IPR Protection: ipr.icegate.gov.in

E-Commerce portal of Central Board of Excise and Customs: http://www.icegate.gov.in

Intellectual Property Appellate Board, India: www.ipab.tn.nic.in

Department of Information Technology, India: www.mit.gov.in

Plant Varieties and Farmers' Rights Authority, India:

www.plantauthority.gov.in

National Biodiversity Authority: http://nbaindia.org/

The Indian IPR Foundation: www.nipo.in

World Trade Organisation: http://www.wto.org

❑❑❑

Chapter 9

Drafting of IP Legal Agreements, Prosecution of Patent Application

P K Ashwini Kumar

Intellectual property has assumed a high degree of importance worldwide and more so in India. This change has spurted critical importance ever since countries became signatory to Trade Related Intellectual Property System (TRIPS) agreement under the aegis of WTO.

Intellectual property cannot be divorced from trade practices within a country as well as across countries worldwide. Businesses in developing countries are very critical of IP protection of any import of goods, technology in their country of interest as they want their business interests to be safeguarded against any infringement or copying of the imported item .

When we talk of Intellectual property, it encompasses these major aspects : Patents, design, trademark, copyrights. There are two other forms of IP : Geographical indicators and trade secrets which may have an altogether different framework in exploitation. Therefore, in this discussion, only the earlier three stated forms of IP will be focused on.

Why legal agreements for IP? Intellectual property is an asset created by innovators , either individually or as a team with or without affiliation to a legal entity. Since it's an asset, it is obvious that one can trade this for economic benefits and hence, safeguards need to be put in place to prevent any unlawful exploitation of these assets. The details of the assets are available in public domain as the contents are published by the state granting the intellectual property rights to the entity creating the IP. This makes it necessary to enter into a legally binding agreement between the parties to be able to exploit the benefits of these assets for mutual benefit.

IP legal agreement is basically a contract to be signed in faith by two parties engaged in trading of the asset. In a very broad sense, this important document will contain: An agreement includes an *offer* and an *acceptance*. Any promises made by parties must be supported by legally-sufficient and bargained-for consideration.

This is something of value that is received or promised to convince a person to make a deal. Both parties entering into the contract must have the contractual capacity to do so; the law must recognize them as being competent. The contract's purpose must be to accomplish some goal that is legal and not against public policy.

Therefore, while an IP legal agreement is being planned to be entered into between two parties, it is necessary to spell out which particular IP is being agreed upon . For example, if a party A is wanting to enter into an agreement to make available the contents of a patent (application filed or patent awarded!!), the said party will be required to spell out clearly its intention to make available the contents to the party B who showed interest in the patent .The party B will be required to clearly spell out that it desires to take the contents from party A. Further the two parties have to put on record in the agreement the financial considerations of payment of premium and royalties which will be a binding on both the parties. One of another aspect of an IP legal agreement is the legal binding on the receiving parties A and B about any further developments which might accrue during the tenure of agreement. The agreement also need to have a definite timeframe and the scope i.e whether exclusive or non exclusive passing of contents and exploitation of the same.

Who should participate in drafting of the legal agreement ? Any IP agreement should be framed jointly by the team of innovators, finance team and legal team. Latter will finally be responsible to approve the final agreement.

Prosecution of Patent Application

Patent as mentioned earlier is one aspect of Intellectual Property and has a very strategic value in any country regardless of whether the IP is created by an individual of group of individuals or any legal entity. Strictly speaking, patent prosecution means the process adopted by any patent office, after it receives the request by any applicant for grant of a patent. The request, being in the form of certain mandatory forms as required by the respective patent office and a detailed description of the invention. In a much broader sense, patent prosecution should mean the entire process of drafting, filing of documents with patent office, attending to the comments of the patent office on the submitted technical documents with claims, effecting acceptance of the complete documents with amendments in the submitted documents leading to grant of patent. This is called the pre-grant prosecution.

The post grant prosecution involves in annual maintenance of the granted patent by paying annuity fees to the respective offices, submitting of licensing agreements to the patent offices, in case the patent is exploited for commercial use. The post grant prosecution could also have the step of non-continuation of the validity of patent by formally informing the patent office to such an effect with concurrence from the patentee. Generally, all the patent offices have almost similar prosecution practices, with just a few differences here and there.

In the present discussions, patent prosecution in the pre-grant situation is touched upon for Indian patent office, The Patent Co-operation Treaty (PCT) system, the European Patent Office and United States Patent Office. The prosecution stage prior to submission of document to any patent office deals with the interaction of the innovators with the expert in the patent law of the country. This is depicted in Fig. 1 below :

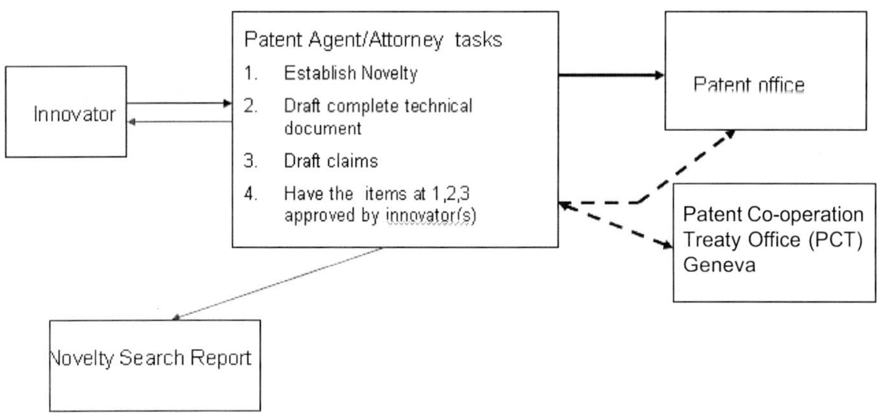

Fig. 1 :

The prosecution of the patent application received by any patent office is shown in Fig 2 below :

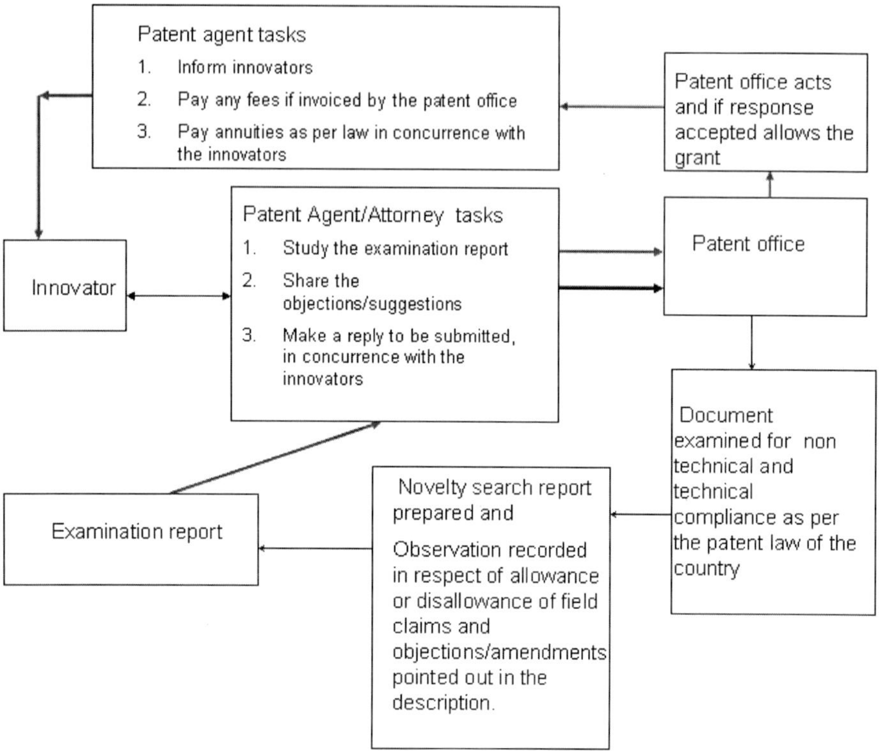

Fig. 2 :

In the patent co-operation treaty system (PCT) very briefly, the PCT office conducts an international search and shares the findings with the agent giving an opinion about existence of novelty of the invention. Based on this, the innovator may amend the application, if novelty aspect is questioned as per the search report. The application after any amendments are submitted to the PCT office within the prescribed time limit. The PCT office then publishes the application after 18 months of first submission of the patent document in any patent office. Then the amended application may be submitted to the desired country of choice as per business interest. However, this still does not warrant grant of the patent in the country of interest. The patent office in question will still go through the law of the country and submit its examination report which will have to be attended to as shown in flow chart in Fig. 1.

References

1. http://www.training-jpo.go.jp/en/uploads/text_vtr/pdf/TRIPs_Agreement.pdf
2. http://ipindia.nic.in/PatentOfficeProcedure/PatentOfficeProcedure_2009.pdf
3. http://www.wipo.int/pct/en/
4. http://www.epo.org/applying.html
5. http://www.uspto.gov/patents/process/index.jsp

❑❑❑

Chapter 10

Enforcement of Intellectual Property Rights, Infringement of Patents and Relief

P K Ashwini Kumar

Intellectual property encompasses these major aspects: Patents, design, trademark, copyrights. Further, Intellectual property is an asset created by innovators , either individually or as a team with or without affiliation to a legal entity. Since its an asset, it is obvious that one can trade this for economic benefits. When IP rights are conferred on an innovator, individual or a legal entity in any state, the IP holder may not necessarily license it for exploitation. However, there is still a need to safeguard against unlawful use of the IP by any entity in that territory. This is essentially what is called as enforcement of Intellectual Property Right, meaning that the holder of the IP will take adequate legal measures to prevent such unlawful use of the asset. The use of an IP without the prior written consent of the asset holder is called Infringement and needs to be tackled effectively to safeguard the business and IP interests of the entity holding the asset.

There is no point in establishing a detailed and comprehensive system for protecting intellectual property rights and disseminating information concerning them, if it is not possible for the right-owners to enforce their rights effectively in a world, where expanding technologies have facilitated infringement of protected rights to a hitherto unprecedented extent. They must be able to take action against infringers in order to prevent further infringement and recover the losses incurred from any actual infringement. They must also be able to call on the state authorities to deal with counterfeits.

Even if there not be any IP protection, yet there is a need to enforce the knowledge generated. The case of India opposing the USPTO granting patent on use of Turmeric powder. Use of turmeric has been there in India from ancient times and still there is no IP protection in India or abroad. Yet another case is of the famous Basmati rice litigation in USA. The US company Rice tech's patent in US was challenged and few claims were disallowed.

Enforcement need not necessarily happen only, when an IP is grated to the applicant. Enforcement can be initiated by any aggrieved party, once the IP becomes available in public domain during the course of prosecution by the IP office of any country. Such a situation arises when any aggrieved party comes to know of the substantially similar nature of the disclosure as his work, even if the party might not have the IP protection. Such a situation helps businesses to prevent erosion of market share where such an entity might not have the IP protected but has a product in the market much earlier than the infringing party. In such a case, the affected party approaches the Intellectual property office of the relevant jurisdiction and files an opposition against the infringer. The possibilities for opposition arise particularly in relation to registrable rights, such as patents and trademarks, because the registration can be disputed. The possibility of opposition rarely arises in relation to copyright and related rights, since these rights in the great majority of jurisdictions arise automatically on the creation of the protected work.

Despite efforts to achieve friendly settlement, circumstances can do arise where the owner of a right feels that he must take action against an infringer in order to protect his markets, present or future. In most jurisdictions, this will be handled in the civil courts.

When thinking about patent enforcement, a patentee must first assess what he has patented. It is now a feature of just about all patent systems that a patent must include either a specification containing claims or a description, claims or any required drawings (depending on the terminology of the particular law). In most systems, the claims are decisive, as they define the scope of protection sought and eventually granted through the patent. The specification or the description and drawings may be used to interpret the claims, which must be fully supported by them.

Once the enforcement proceedings get initiated in a civil courts following may happen. Cases involving complex technology can take a very long time in court with expert witnesses being called and subjected to examination and cross-examination by both sides. In issuing his reasoned decision, the judge can make various awards. In addition to injunctions, the court may also award damages in respect of the infringement, namely compensation for sales and markets lost as a result of the infringing activity. As an alternative to damages, the owner of the patent can be awarded an account of profits. All the profits derived by the defendant as a result of the infringement may be surrendered to the patent owner. An order requiring the defendant to deliver upto the patent owner, or to destroy, any products or articles incorporating the patented invention can also be made. Finally, the patent owner can secure a declaration that the patent is valid and is infringed.

The remedies typically available in intellectual property infringement actions are injunctions, damages and account of profits. Most actions start with an application for some form of preliminary or inter-locutory relief, and in most cases, do not get beyond this preliminary stage.

References

1. www.wipo.int/about-ip/en/iprm/pdf/ch4.pdf
2. http://www.wto.org/english/docs_e/legal_e/27-trips_05_e.htm
3. http://ec.europa.eu/internal_market/iprenforcement/index_en.htm

❑❑❑

Chapter 11

Strategic Use of Intellectual Property Rights in Livestock Sector for Value Extraction

K P Ramesha and Mukund A Kataktalware

In the present knowledge based global economy, protection of intellectual property (IP) and knowledge management are critical for technology development and diffusion in all fields including livestock sector. A paradigm shift is required in livestock sector from production orientation to quality and cost orientation in the WTO era. The knowledge based technological revolution in livestock sector helped dominant players to corner the benefits at the cost of smallholders particularly in poultry and pig industry in many parts of the world.

Livestock Sector and IPR Regime

India possesses one of the largest livestock populations in the world, which is more than 484.9 million. As per the Basic Animal Husbandry Statistics (BAHS), 2012, in India, there are about 56.7% of world's buffaloes (105.05 million) with 1st position, 12.5% cattle (199.07 million) with 2nd position, 20.4% small ruminants comprising 144 million goats and 72 million sheep with 2nd and 3rd position respectively in the world. The value of output from livestock sector is Rs 459,051 crores during 2011-2012, which is 24.8 per cent of total agricultural sector and 25.6 at constant price 2004-2005. The productivity of Indian livestock is low. Technology driven productivity enhancement in low producing animals is the need of the hour. The future development of livestock sector depends on adoption of new and innovative production and processing technologies, and proper market tie-ups. Diversification and high value produce will add new dimensions to this sector.

A paradigm shift is required in livestock sector from production orientation to quality and cost orientation in the WTO era (Ramesha 2011), which could be achieved by technological advancement for which innovation is the key. In the present knowledge based global economy, protection of intellectual property (IP) is critical for technology development and diffusion in all fields including livestock sector (Ramesha *et al.*, 2010). Intellectual Property (IP) is a category of property that confers rights over intangible creations of human intellect. The livestock production takes place within the national boundaries as per customary laws and knowledge; whereas the livestock products are marketed at global level. Therefore, national and international as well as regional laws and Intellectual Property (IP) laws and global commitments are relevant to livestock sector.

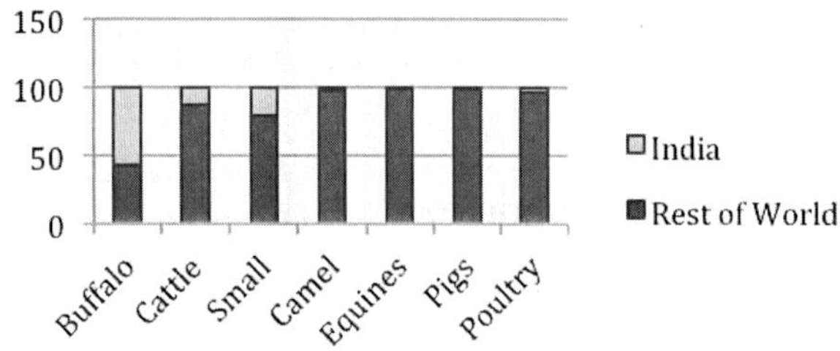

Fig. 1. Proportion of various species of livestock in India in comparison to world population

A large number of knowledge based technologies have been generated till now for enhancing the livestock productivity and in improving the health and reproductive status of the livestock. Livestock products and process technologies will provide higher market through value addition. Knowledge management including their protection in the modern IPR regime is the key to develop technologies in livestock sector. Some of the areas knowledge management have major influence in technology development include animal production technologies including genomic selection, animal health technologies mainly diagnostics and vaccine development; feed technologies; modern reproductive technologies; commercial production of livestock and value addition to livestock produce; livestock waste management; commercial production of probiotics and prebiotics; organic waste management.

Intellectual Property (IP) is a category of property that confers rights over intangible creations of human intellect. IPRs are legal and institutional devices to protect creations of the mind. India is signatory to both CBD and TRIPs Agreement and also ratified Global Plan of Action on Animal Genetic Resources adopted in September 2007, therefore, it is obligatory for India to develop suitable legislation to protect and improve livestock biodiversity and associated traditional knowledge. As per Trade Related Intellectual Property Rights (TRIPs) agreement, any country ratifying the Global Agreement on Trade and Tariffs (GATT) and becoming a member of the World Trade Organization (WTO) must establish minimum standards for intellectual property rights (IPR). TRIPs agreement provides effective protection to following 7 forms of IPRs :

1. Patents/protection of plant varieties and plant breeders rights
2. Copyrights and related rights
3. Trademarks
4. Industrial designs
5. Lay out Designs (topographies) of Integrated Circuits
6. Geographical Indications
7. Trade Secrets.

IPR are largely territorial in nature except copyright, which is global in nature. These rights have to be renewed from time to time for keeping them in force except in case of copyright and trade secrets. The various national and regional patent systems are receiving and granting patents in the field of livestock sector, while policies and regulations concerning the sector are being articulated as a topic for discussion in the CBD and FAO forum. Intellectual property laws have the potential to alter the existing legal conditions for competition and investments in the livestock sector, and needs therefore, a higher level of awareness among policy-makers, breeders, scientists and farmers.

Patenting of living organisms is a contentious issue among the WTO member states. Under TRIPS Article 27.3, members must provide various forms of intellectual property, many of which are relevant to livestock sector, including indications of geographical origins, trademark, trade secrets, patents and copyright. Provision is made for plant variety protection under patents or a *sui generis* system, or a combination of both. There is no comparable system for animals.

Biotechnology firms are interested in patenting processes and genes used in animal breeding as such. The patenting of broad traits like increased milk yield or broad claims on genes/animals may lead to monoculture and monopoly which would result in suffering to the pastoralists, small and marginal farmers. With the advent of genetic engineering and multilateral IPR agreements a new era of gene hunting has started. Researchers, innovators and companies have now started to protect their intellectual property in livestock sector. This has alarmed and threatened the general public and raised economic, legal and ethical concerns (Rothschild *et al.*, 2003).

Patenting in Livestock Sector

A *Patent* is a statutory right for an invention granted for a limited period of time to the patentee by the Government, in exchange of full disclosure of his invention for excluding others, from making, using, selling, importing the patented product or process for producing that product for those purposes without his consent. Any product or process or design which involves an inventive, novel and non-obvious step capable of industrial application could be patented. An invention relating either to a product or process that is new, involving inventive step and capable of industrial application can be patented. However, under Indian Patent Act, it must not fall into the categories of inventions that are non-patentable under section 3 and 4 of the Act. The aim of a patent system is to encourage economic and technological development by rewarding intellectual creativity. Intellectual property system safeguards the rights of an inventor in his invention/intellectual richness while benefitting the end users and society as a whole. The purpose of patent system is to protect inventions and encourage fair competition in the field of technology so as to transform inventions or creations into real and productive forces. It is to be remembered that the possession of a patent not only confers certain monopoly rights and privileges of the patented article, but certain obligations and duties also. Commercialization of IP-enabled technologies is an absolute necessity to transfer valuable technologies developed in the institute laboratories to the society for greater benefit of mankind.

Under the TRIPS agreement, farmers have to get legal protection for their traditional knowledge via patents or by an effective *sui generis* system or by both. National Bureau of Animal Genetic Resources

(NBAGR) has been identified as a nodal institute to develop a system to register and document valuable animal genetic resources by ICAR. The registration system will bring elite germplasm into public domain to promote its use in research, however, the system does not have any legal backing. In India, in the recent years there is a study increase in the number of patent applications filled in the area of animal and Veterinary Sciences. Good number of patent applications has been filed by various ICAR institutes and agricultural and allied sector universities in the field of animal, veterinary sector.

Biotechnology is providing the capability to produce novel genetically modified organisms (plants, micro-organisms, fishes and animals), which are commercially valuable and industrially useful. In the livestock sector, patents have been granted for gene sequences with utility, such as genetic markers. A New Zealand company, Agmark, has claimed a patent on the "Booroola" gene, which regulates the ovulation rate in sheep. The Booroola gene can be traced back to Bengal sheep, which were imported from Kolkata and crossed with Merinos (Köhler-Rollefson, 2005). The patent covers animals that are produced in a breeding programme, in which the DNA test has been used, but not those animals that carry the gene naturally. The open question is whether the patent also covers the offspring of the animals that have been tested for the presence of the gene. Another notable patent application is one by Monsanto for a series of twelve patents on pig breeding. One of the first conflicts in quantitative animal breeding resulted from a patent entitled "Method of Bovine Herd Management" granted to the Cornell Research Foundation (Cornell University) in the USA, in 1994 and Canada in 1998. The invention is for the "test-day model" and includes the gathering of data, mathematical treatment and the use of the data by dairy producers (Rothschild *et al.*, 2003; Scholtz and Mamabolo, 2006). The novelty and non-obviousness of the patent has been seriously questioned. The practices of gathering, manipulating and using data by dairy producers have existed for nearly 100 years (Schaeffer, 2002). In South Africa, these principles have been in use since 1917. The patent, therefore claims rights to a practice that has been public knowledge for a long time. The novel idea within the patent is the specific mathematical model and procedures that were developed for the analysis of test day yields.

In European Union and some other countries, patents can also be granted on normal animals, which have merely been subjected to certain techniques like a gene diagnosis, or a process for determining the animal's sex. The European Patent Office granted patent on Dolly the cloned sheep to the Roslin Institute in Edinburgh (patent application EP 849 990) in 2001. The US Company XY Inc. was in 2005 granted patent EP 1257168, which covers a method for selecting sperms by sex for the artificial insemination of mammals. The first European patent on genetically manipulated dairy cows was granted in 2007. Under patent number EP 1330552 "inventors" from Belgium and New Zealand claimed processes for breeding cows, which give more milk or milk with altered constituents. The cows are produced either by marker assisted breeding and bred normally, or by having more milk genes additionally incorporated into their genome. Hungary is one of the few countries that grant patent protection for animal breeds (WIPO, 1995). The Bulgarian Law for Protection of New Plant Varieties and Animal Breeds has been enforced since 1996. In terms of patentability of higher life forms, such as animals and fishes, they are not treated any differently by the Australian Patent Office from the way in which lower life forms are treated. The East African Boran cattle breed has also been patented in Australia (Scholtz and Mamabolo, 2006). The practice in New Zealand is similar to that of the Australian Patent Office. In general, patents to animals are also allowable in Japan. In contrast to these countries, animals (whether transgenic or otherwise) are not patentable in countries, such as Argentina, Brazil and China (Blattman *et al.*, 2002) and India (Ramesha *et al.*, 2010, 2011).

Patenting of Livestock Breeds

Hungary is one of the few countries that grant patent protection for animal breeds (WIPO, 1995). The Bulgarian Law for Protection of New Plant Varieties and Animal Breeds has been enforced since 1996. In terms of patentability of higher life forms, such as animals, they are not treated any differently by the Australian Patent Office from the way in which lower life forms are treated. The East African Boran cattle breed has also been patented in Australia (Scholtz and Mamabolo, 2006). The practice in New Zealand is similar to that of the Australian Patent Office. In general, patents to animals are also allowable in Japan. In contrast to these countries, animals (whether transgenic or otherwise) are not patentable in countries, such as Argentina, Brazil and China (Blattman *et al.*, 2002) and India (Ramesha *et al.*, 2010, 2011).

Trademarks

Trademarks are signs or symbols registered by a manufacturer/ producer or merchant to identify goods and services. A trademark can be a word, phrase, symbol, design, or a combination thereof. Many countries also recognize "collective marks" and "certification marks". Collective Marks are a sign which distinguish the origin of a product, method of production and distinctive characteristics of goods or services. A collective mark may be owned by an institution or co-operative or a public entity and used by members of the organization, example- "AAA". Certification marks - like CE (European conformity) and The Good-housekeeping seal of approval - "certify" that products or services meet a particular standard of quality, regional origin or the like, even though the products may come from various unrelated companies. Certification marks-are a sign used to denote the origin of the product that meets set standards and specifications. Certification marks are similar to collective marks except that users of certification marks should not necessarily be members of the organization that owns the mark, example- "Wool Mark". Genetic material *per se* cannot be protected or covered by a trademark. A trademark can be a useful value addition tool. The trademark can improve the value of a product. For example, semen straws, dairy products from a reputed company with trademark may fetch higher price.

Geographical Indications

Geographical Indications (GIs) are signs or expressions used on goods to indicate that a product or service originates in a country/ region or specific place where a given quality, reputation or other characteristic of the good is essentially attributable to its geographical origin. Both Trademarks and GIs are used to identify the products and help to create wealth and improve competitive position of businesses. A Trademark puts emphasis on the producer or a manufacturer of a product, whereas a GI highlights the geographical origin of a product and the characteristics derived from it. GI does not protect the breed or genetic material *per se* but add commercial value to the animals of a breed reared in a particular region. They may be used as a value addition mechanism. Geographical indications prevent all others other than the producers of goods from the original region from using that particular geographical indication incorrectly as a trademark for other products than those from the area. Registration of Darjeeling tea-cultivated in 87 designated locations in the Darjeeling

Region helped to prevent misappropriation and improved marketing position and income.

Ethiopia produces some of the finest coffee in the world but gained very little as there was a problem of misappropriation, but filing of trademark applications in 36 countries and securing protection in 30 countries helped to differentiate the Ethiopian fine coffee from the coffee of other countries. This helped to build confidence and improve bargaining position of coffee producers and exporters from Ethiopia and increased the price of Ethiopian coffee and income of coffee producers, which resulted in a positive improvement in the living standard of farmers. In France, Bresse breed of chicken is produced in the Bresse geographical area only, as defined by law under protected designation of origin (Verrier *et al.*, 2005) and sold at 50-60 per cent higher price. Similarly, to save the Sambucana sheep from extinction, a consortium was set up and a special brand name for guaranteed Sambucana meat was set up in Italy. Roquefort cheese can be made only from the milk of the Lacaune sheep breed. Chilika buffaloes with typical curd made from its milk, and Banni, Dharwari buffaloes with typically flavoured milk products are other examples, which could be considered for protection (Ramesha, 2008). Under GI, the production system is characterized by natural conditions and a production system unique for that area. There is scope for utilizing GI based on unique characteristics of products produced from a breed in a geographical area due to natural conditions and unique production system.

Trade Secrets

Trade Secrets (in certain circumstances and jurisdictions called know-how) provide protection against unauthorized disclosure or use of confidential information (of a technical or commercial nature). To be protected as a trade secret, the intellectual property must, of course, be kept secret, and must also coinfer some sort of commercial advantage to the holder. Enforcement of IP rights for trade secrets is possible, when a competitor has misappropriated and/or stolen the trade secret. Trade secrets are important in animal and fish breeding where commercial breeders want to keep their nucleus stock and the pedigree information (and other information) away from their competitors. A point often raised is when one should file for a patent or maintain the information as a trade secret. What is important to note is that patents and trade secrets are not in conflict with each other but are complimentary IP assets. Depending upon the nature of the know-how; or the invention, the organization may choose to either file a patent or to continue to hold as a trade secret.

IPR issues in the field of livestock sector

- Before the availability of modern biotechnological tools, inventions based on living organisms were treated as natural phenomena i.e discoveries and were not patentable.

There are concerns in the area of :

- The limits of patentability

  In biotechnology research, many times it is difficult to distinguish between invention and discovery. The isolation of a gene from its natural environment and the identification of its function render the gene and its sequence an invention for patenting purposes in some countries.

- Patenting of "enabling technologies".

  This issue is of great importance as it has an impact on access to these technologies by the livestock research system in general.

- The multiplicity of patents required to develop a livestock product.

  Developments in modern livestock biotechnology require the use of several processes and products, which in most cases will be subject to patent protection. For example, producing transgenic livestock involves several processes such as identification of gene of interest, inserting the gene at appropriate stage, to check the gene incorporation, to study the gene expression, checking the animals having inserted genes, etc.

- Patents on specific genes usually extend to the GM organisms into which the genes are inserted, thus bringing the entire organism under patent protection.

  This question has raised considerable debate. Eventual financial revenues are granted to the patent holder, without compensating the developers of the original plant/animal variety.

- Concentration of the IP in the hands of few MNCs.

  In the modern biotechnology era, there is danger of animal genetic material from biodiversity rich countries being exploited by others including large international companies. Another important element is that a small number of multinational companies (MNCs) dominate livestock biotechnology field. Existence of strong IPRs, and the fact that the rights over the technologies and products are often owned by MNCs, might lead to (increased) dependence by developing country farmers on MNCs and developed countries. With the development of modern biotechnology, there is possibility that indigenous breeds/varieties may be genetically altered and the new variety/breed may later be substituted for the original breed/variety from which it was developed.

- Patenting of indigenous livestock genetic resources.

  Patents can and have been issued to companies from developed countries over genetic material from developing countries. In some cases, the lack of appropriate mechanisms for sharing of benefits has generated considerable controversy.

Need of the hour

The development of a legally binding "International Framework on Animal Genetic Resources" is, therefore, becoming a matter of urgency. In the modern IPR era, it is essential to provide certain rights such as access to natural resources, grazing rights, participatory decision, making in the plans or programmes affecting ANGR to livestock keepers through a sui generis system which will help in protecting native livestock and their sustainable use. Both the herd registration scheme and breed registration system developed by ICAR lack legal enforcement authority. There is an urgent need to develop a suitable legally binding *sui generis system* and establishment of "**Animal Genetic Resources Authority of India"** for protecting the interests of livestock keepers and animal breeders through a law comparable to the Protection of Plant Varieties and Farmers' Rights Act, 2001 and PVP and FR Rules 2003 with due consideration for the special characteristics of AnGR (Ramesha *et al.*, 2007, 2010).

Epilogue

Protection of native livestock breeds/varieties and associated traditional knowledge through appropriate *sui-generis system* is essential for livestock improvement in the modern IPR regime. Identification of genes for unique characteristics in native livestock breeds will go long way not only in the advancement of science and livestock production, but also pave way for patenting of gene sequences for these traits. If proper internationally accepted, legal system is developed with respect to use of livestock biodiversity, patenting of novel genes will bring about economic benefit to the livestock keepers in the modern IPR regime through benefit sharing and help in conservation and sustainable use of native livestock breeds. The IPR and other legal instruments influencing knowledge intensity of livestock sector is complex and are greatly influenced by knowledge generation and management. Our country with large number of well trained veterinary and animal science personnel and large livestock resource should develop a suitable environment for innovation by appropriate Science and Technology policy as well as IP policy, which is critical for economic development of the sector which is concerned with livelihood security of huge number of rural folk. The IPR and other legal instruments influencing knowledge intensity of livestock farming, access and transfer of livestock genetic resources and livestock produce is complex and are greatly influenced by knowledge generation and management. We should make use of all the available flexibilities in TRIPS in IP system to suit the specific needs particularly with respect to traditional knowledge. Through strategic use of Intellectual Property Rights in livestock sector, it is possible to extract proper value to livestock and its produce.

Note: Some portions of this article has been taken from the earlier presentations of the author.

References

Blattman, J., McCann, J., Bodking, C. & Naumsoka, J., (2002). Global intellectual property : International developments in animal patents. In: Intellectual Property Rights in Animal Breeding and Genetics. Eds. Rothschild, M.F. & Newman, S., CABI Publishing, New York. pp. 63-84.

Köhler-Rollefson, (2005). Building an International Legal Framework on Animal Genetic Resources can it helps the drylands and food-insecure countries League for Pastoral People.

Ramesha K.P., Pourouchottamane R and Bhattacharya, M. (2007). Role of indigenous livestock genetic resources in the era of modern Intellectual Property Rights. Indian Dairyman, 2007, 59 (11): 39-45.

Ramesha, K. P., Sadana, D. K.and Goswami, S.L. (2010). Modern IPR Regime- a boon or a bane for livestock biodiversity. Indian Dairyman. 62(8): 74-79.

Ramesha, K.P. (2011). Intellectual Property Rights Regime for livestock agriculture in India - present status and future prospects. JIPR. 16(2): 154-162.

Rothschild, M.F., Plastow, G. & Newman, S. (2003). Patenting in animal breeding and genetics, in A. Rosati (ed.) *WAAP Book of the Year 2003*. Wageningen Press, for World Association for Animal Production (WAAP).

Schaeffer, L.R., (2002). Dairy cattle test day models: A case study. *In: Intellectual Property Rights in Animal Breeding and Genetics*. Eds. Rothschild, M.F. & Newman, S., CABI Publishing, New York. pp. 233-246.

Scholtz, M.M. Mamabolo, J. (2006). A developing country perspective on recent developments in animal breeders and intellectual property rights. African J. of Biotechnology. 36(5): 22-25.

Verrier, E., Tixier-Boichard, M., Bernigaud, R. and Naves, M. (2005). Conservation and value of local livestock breeds: usefulness of niche products and/or adaptation to specific environments. Animal Genetic Resources Information. 36: 21–31.

WIPO (1995). Database of Intellectual Property, Hungary Law No. XXXIII of 1995 on the Protection of Inventions by Patents, Pp 1-35.

□□□

Chapter 12

Technology Pricing and Valuation in the Intellectual Property Rights Era with Special Reference to Livestock Sector

K.P. Ramesha, Letha Devi G. and M.A. Kataktalware

India possesses one of the largest livestock populations in the world, which is more than 484.9 million. The total livestock of the country comprised of 37.6% cattle, 19.9% buffalo, 13.5% sheep, 26.5% goats and 2.1% other livestock species. The productivity of Indian livestock is low. Technology driven productivity enhancement in low producing animals is the need of the hour. In livestock sector, it is essential to have multipronged strategy involving technology generation through research, technology dissemination, technology users, and supporting mechanisms (input supply, market credit) and quality improvement through technological interventions for productivity enhancement and to manufacture quality products.

Technology Management

Technology can be defined as the creation, usage and knowledge of tools, techniques, crafts, systems or methods of organization to solve a problem or serve some purpose. Technology Management is the management of innovation whether it is a product, a process, or an organization from its conception to its diffusion and therefore, to its implementation within the organisation, including the consequences, advantages and disadvantages for all variables or actors involved in running the organisation. Technology commercialization is the process of converting knowledge into products and services. It is aimed at translating knowledge gained through research for the benefit of the society and is essential for the socio-economic development of the society.

The development of technology can be viewed from various perspectives. It can be viewed as:

- An engine for economic development,
- A strategic resource, and/or
- A competitive weapon

Therefore technology management assumes an important role right from activities related to planning and developing the technological capabilities of an organization. Active technology management usually occurs in the following three ways:

i) Ensuring that a organisation's competitive advantage is optimized through good use of existing IP and new filings

ii) Generating additional revenue-creating opportunities through licensing activities

iii) Generating additional business advantages by licensing in relevant IP.

Technology Generation in Livestock Sector

A paradigm shift is required in livestock sector from production orientation to quality and cost orientation in the WTO era (Ramesha, 2011), which could be achieved by technological advancement for which innovation is the key. In the present knowledge based global economy, protection of intellectual property (IP) is critical for technology development and diffusion in all fields including livestock sector (Ramesha *et al.*, 2010). In the WTO era, a vast amount of opportunities have emerged for the R & D organizations, entrepreneurs as well as for the professionals engaged in livestock sector. Apart from rearing of livestock and poultry and livestock based industry has vast scope for entrepreneurship development. In India, in the field of animal sciences many technologies are generated by the public funded institutions. However, technology commercialization and entrepreneurship development is very less. ICAR institutes and State Agricultural Universities and NGOs have created knowledge in animal sector that has intrinsic intellectual property with potentially significant benefits to the society.

Livestock products and process technologies will provide higher market through value addition. Diseases reduce the production potential and cause great economic loss to the farmers in the form of qualitative and quantitative drop in produce. The IPR has a bearing on the development of biological products (vaccines) and diagnostics. IPR laws also have a bearing on the development and diffusion of technologies in the field of production as well as in processing. It has a major role to play in technology development and diffusion aimed at value addition to livestock produce. In the years to come, the biosensor system are likely to aid farmers in taking precise decisions pertaining to breeding, feeding and healthcare. The present IP regime in the country is likely to enhance technology transfer from developed countries to India in the area of greener technologies aimed at reducing methane production from ruminants and quality improvement in dairy products through technological interventions. The traditional knowledge in the sector could be used effectively for bringing benefit to India in the form of benefit shearing in the IP regime.

Technology pricing is important for

i) Licensing (commercial, further R&D)

ii) Assignment

iii) Partnership

iv) Joint venture

v) Launch of your own company

vi) Mergers/takeovers

vii) Enhancing market value

viii) Raising funds

Valuation of Government Funded Projects

Normal tendency is to do the valuation around the sanctioned amount for the project which is not the correct valuation of an IP as by the time you are ready to license the IP, you have spent many years of research and hard work. There is an opportunity cost due to depreciation of equipments, buildings, cost towards protecting IP and so on.

Technology evaluation

Technology evaluation could be Qualitative valuation or quantitative valuation. Qualitative valuation focuses on the worth of IP due to its particular features and methods of exploitation. Qualitative valuation is generally done by PRISM method and IP is classified into four basic categories viz.:

- Monopoly IP-which are of high value
- Defensive IPs-are generally are of low value
- License-high value
- Joint Venture

For selling or licensing of a technology, it is essential to know the quantitative valuation of a technology. It is impossible to assess the total value of unproven, new technology. For financial valuation of technologies, the following approaches are being used for quantitative valuation.

i) Costs approach: The pricing of a product is based on the cost of developing the product.

ii) Income approach: The value of a technology is determined by a pure income approach, whereby future anticipated revenues (cash flows) are discounted to present value.

iii) Market approach: The value of a technology is determined based on the value of similar or comparable technology.

iv) Hybrid approach: The value of a technology is determined by a combination of income and market approaches.

v) Royalties approach: The value of a technology is calculated based on royalty rates that have been applied to similar technologies.

Avenues for extracting value for IP : The following avenues are available for extracting value from IP and IP enabled technologies:

- Selling IP Rights;
- Licensing
- Engaging product development company
- Developing product or company

- Alone utilizing IP
- By contract for services
- Through corporate partner

Extracting Value from IP through Licensing

Extracting value from IP through licensing is commonly used by R & D institutions. Patent licenses and technology licensing agreements allow one party to further develop, use, make, or sell the patented (and/or trade-secret-protected) technology of another party. With a license possession of property remains with the original owner. Patent licenses may be specific to one or several patents. Technology licenses usually include the transfer of know-how (which may or may not be a trade secret) and sometimes, materials. Such license may provide for the further development of a technology or limit production/ manufacture of a good or provision of a service. The goal of a negotiation is to formulate an agreement that meets the needs of both the licensor and the licensee in a manner that ensures mutually beneficial future relationships between institutions and individuals. A license agreement will include several key components (Daren Bolding, 2011):

- The background section sets out the factual predicates for the license.
- The definitions section explains key terms used in the agreement.
- The grant section establishes whether only the licensee may practice the invention (an exclusive license) or whether others may practice it.
- The section on fixed payments and royalties sets out payment terms of the agreement.
- The confidentiality section specifies provisions for and restrictions on the disclosure of information shared between the parties.
- A section on enforcement against infringers specifies which party shoulders the burdens and realizes the benefits of enforcing the licensed patent against infringers.
- The term of the agreement and termination procedures should be fully spelled out.

As a practical matter, patent licenses are most valuable, when coupled with access to associated know, how. Patent applications are usually filed early during the research stage to get the earliest possible filing or priority date. The best mode for commercial manufacture and use are almost invariably developed later. A patent license alone is, therefore, often inadequate for commercial development of the technology. In licensing complex technologies, the licensor can partition patent rights based on duration, location and field-of-use. By partitioning a bundle of patent rights and distributing them to one or more licensees, field-of-use licensing maximizes value, optimizes delivery, and facilitates the most effective use of new technologies. The result can be multiple licensees, each with different and specialized access to the technology; which can efficiently speed different types of products to market. By limiting licensees to a particular field, a licensor retains the ability to work with the best possible licensee(s) for a new use, when it arises.

Licenses that include royalty stacking and royalty packing clauses are becoming more common because virtually all products developed now using biotechnology, genetic engineering, and chemistry are technologically complex and incorporate many different inputs (Daren Bolding). There are two scenarios that can arise, when multiple royalty rates are attached to one product–royalty stacking and royalty packing:

- Royalty Stacking. A biotechnological product might have multiple patents attached, and thus require multiple licenses in order to make, use, or sell the product.
- Royalty Packing. With some biotechnologies, it usually is necessary to combine one technology with many other technologies. (In this situation, the royalty imposed on each of the proprietary products are administered will be packed together).

Commercialization of IP and Entrepreneurship Development

Entrepreneurship is the process of starting, establishing and enlarging an enterprise that produces and markets goods and/or services. An entrepreneur identifies opportunities and seizes opportunities for economic benefits. According to Cole "Entrepreneurship is the purposeful activity of an individual or a group of associated individuals undertaken to initiate, maintain and enhance profit by production or distribution of economic goods and services".

It involves an attitude that tries to find opportunities and then derive benefits from them. Entrepreneurship is an important process by which knowledge is converted into products and services. The process of Business Incubation for Entrepreneurship (BIE) is a critical organizational support mechanism for fresh entrepreneurs at the initial stage. A typical high quality Business Incubator provides the essential services to a budding entrepreneur, which includes physical infrastructure, administrative support, management guidance/ mentoring, help in formulation of a business plan, technical support, Intellectual Property(IP) advice where applicable, facilitating access to finance and encouraging networking with the greater and relevant business community. At the start-up stage, the entrepreneur is beset with significant challenges of marketability and resources (financial and otherwise) which successful incubation can help to address. An entrepreneur not only perceives the business opportunities but also mobilizes the other resources like 5 Ms-man, money, machine, materials and methods.

In countries like India, where the biotechnology industry is in the early stages of development and where there is a smaller chance of finding an acceptable licensee, creating a spinout company may hold more promise. Universities and public sector research institutes need to be relied upon as sources for human capital and investment in entrepreneurship. Research institutions and government policies are able to constructively influence the creation of new companies. Favourable environment for creating and growing new companies consists of an encouraging business culture (one that rewards success and treats failure as a learning opportunity), access to intellectual capital (such as that flowing from universities and public research institutions), sufficient financial capital, and reliable physical capital (facilities, laboratories, communications). Governments can encourage entrepreneurship by providing a favourable environment. Once enough companies exist, they will themselves further transform the environment, attracting or creating the skills and capital that can develop into a technology cluster.

Scope for Entrepreneurship Development in Livestock Sector

The future development of livestock sector depends on adoption of new and innovative production and processing technologies, and proper market tie-ups. Diversification and high value produce will add new dimensions to this sector. Proper post-harvest handling, reduction of losses and hygienic primary processing are important to realize full potentials of these sectors. Vast scope exists in the following areas for entrepreneurship development:

- Technologies those reduce the cost of production and increase the benefit of the farmers.
- New technologies that are simple and time saving and keep away farmers from drudgery of labour.
- Technologies that provide social and psychological benefits to farmers.
- Processing units and agro produce manufacturing units.
- Inputs manufacturing units: These units produce goods either for mechanization or for increasing productivity.
- Service centres : These include the workshops and service centres for repairing and serving the implements, Supply Chain Management-Supply Chain Management (SCM) represents the management of the entire set of production, manufacturing/ transformations, distribution and marketing activities by which a consumer is supplied with a desired product.

Epilogue

It is absolutely essential to commercialise IP-enabled technologies developed in laboratories to the society for greater benefit of mankind.The IPR and other legal instruments influencing knowledge intensity of livestock sector is complex and are greatly influenced by knowledge generation and management.

Note: Some portions of this article has been taken from the earlier articles of the first author.

References

Parente, Stephen L. and Prescott. E.C. 2002. Barriers to riches. Cambridge, MA: MTP press.

WIPO. 2005. Making intellectual property work for developing countries. Submitted to the third session of the WIPO Inter-sessional Inter-governmental Meeting on Proposals for a Development Agenda for WIPO, 20-22 July 2005.

Ramesha, K.P., Sadana, D.K. and Goswami, S.L. (2010). Modern IPR regime-a boon or a bane?. Indian Dairyman. 62(8): 74-79.

Ramesha, K.P. (2011). Intellectual Property Rights Regime for livestock agriculture in India-present status and future prospects. JIPR. 16(2): 154-162.

Technology Management. PGD-TMA2013 Resource Material. NAARM & UoH, ISBN No. 2218 pages 298.

❑❑❑

Chapter 13

Intellectual Property and Technology Management in ICAR Scenario: Need for Intellectual Property Valuation in R&D

C N Ravishankar

The Agricultural Research Institutions under the Indian Council of Agricultural Research (ICAR) plays a lead role in spearheading the scientific and technological development of the country. With the advancements made by the information technology and the increasing use of computers for the efficient management and dissemination of knowledge, the information systems began to play a vital role in agricultural planning and development. As the importance of databases and their management in agriculture is well-recognized and as information processing has become the basis of a good decision support system, there arouse a need for finding an ideal solution for creating a well-crafted data management system that gives the power to efficiently direct the intellectual assets of an R&D organization.

India as a member of the World Trade Organization (WTO) is obliged to comply with the Agreement on Trade Related Aspects of Intellectual Property Rights (TRIPS Agreement),which requires since 1 January, 1995 that member countries provide for intellectual property rights (IPR) in one form or the other in all fields of technology, including agriculture. The transfer of IPR enabled agricultural technologies through commercial route gained greater importance. In response to the changing scenario of technology generation and dissemination, the ICAR, developed a policy framework that will guide the management of IP created by its scientists/innovators at its institutions or elsewhere, and that developed with its support (ICAR, 2006). ICAR moved a way forward in building an understanding among the Scientists and Institutes as to what is going to be the future business of a public research organization under the changing trade and genetic resources regime.

We have been witnessing a steady improvement in IPR and related activities in ICAR. In this IP regime, the Council has made significant contributions through technical-backstopping and inputs to the development/ amendment of new national legislations on plant varieties, patents, other IPRs, biodiversity, seeds, etc, besides our input on several other related dimensions in the area domain of other Govt. of India Departments. ICAR has marched ahead in generating technologies towards sustainable advancements in agriculture and allied sectors, improving quality of higher agricultural education, institutional capacity building, fostering international linkages for co-operation in agricultural research. The research and development activities during the year covered wide areas, ranging from optimizing resources' use, improved cultivation techniques, development of improved varieties/breeds, excellence in agricultural education and frontline extension of technologies, providing improved planting materials and imparting training to various stakeholders.

Efforts undertaken from ICAR Headquarters

As per the ICAR guidelines, the Central Technology Management Committee (CTMC; short title for Central IP Management and Technology Transfer/Commercialization Committee) at the ICAR headquarters, chaired by Director General, ICAR was made the apex decision-making body. All matters of policy concerning IPR portfolio management and technology transfer/commercialization are decided by the CTMC. At the respective ICAR institutions, the Institute Technology Management Committee (ITMC; short title for Institute IP Management and Technology Transfer/Commercialization Committee), chaired by the Director of the Institution, is the final decision-making body for IP related matters/progress/concerns. At the zonal level, the Zonal Institute Technology Management Committee (ZITMC; short title for Zonal Institute IP Management and Technology Transfer/Commercialization Committee) was designated to take decisions for the IPR portfolio management and technology transfer/commercialization of the ICAR institutions in the zone and also the inter-institutional matters. The ZITMCs also advise the institutions in the zone regarding their IPR portfolio and technology transfer matters.

These central, zonal and institute level committees will take steps to coordinate, harmonize and synergize with other relevant committees at the ICAR headquarters/institutions. At the level of ICAR institutions, Institute Technology Management Units (ITMUs) are established. These units are responsible for IP protection/management and technology transfer/commercialization with internal capabilities as well as external legal and business experts, wherever required. The ITMUs acts as Secretariat for the respective ITMCs. Selected ITMUs at national/central institutions in different zones shall be designated as Zonal Agro-Technology Management Centres (ZTMCs). These zonal centres will be strengthened for act as Secretariat for the respective ZITMCs. They advise and coordinate the IPR portfolio and technology transfer matters in various ICAR institutions in their respective zones. The ITMUs and ZTMCs follows the guidelines and policy decisions taken by CTMC from time to time. The ATMC oversees and provides necessary advice and support.

The ICAR's intellectual property (IP) portfolio has been firming up since the implementation of its management guidelines *w.e.f.* 2 October, 2006. The three-tier IP management mechanism implemented in ICAR is bringing commercial ethos in transfer of skills and products. The new Plan Scheme implemented, is a unique in Public-Public Partnership mode (having total project outlay of Rs 486 million in XI Plan), which has a contribution (in 2:1:1 ratio) from the Central Government (Plan Funds), the National Agriculture Innovation Project, and the ICAR's internal resource generation. This Intellectual Property Management and Transfer Commercialization of Agricultural Technology Scheme broadly aims at up-scaling of existing component of IPR previously handled under the ICAR (Hq) Scheme on Management of Information Services. The 3-tier institutional mechanism is focused on protecting/managing the IP generated within the ICAR system, including that in the collaborative research.

Accordingly, at the base level, 95 Institute Technology Management Units have been established and empowered by the appropriate delegation of powers to enhance the institutional IP management capacity. In the middle-tier, 5 Zonal Technology Management & Business Planning and Development Units were established at the identified ICAR institutes in different zones for building capacity for business planning and development in association with NAIP. The zonal centres and the central IPR and Policy establishment help the Institute Technology Management Unit on a case to case basis.

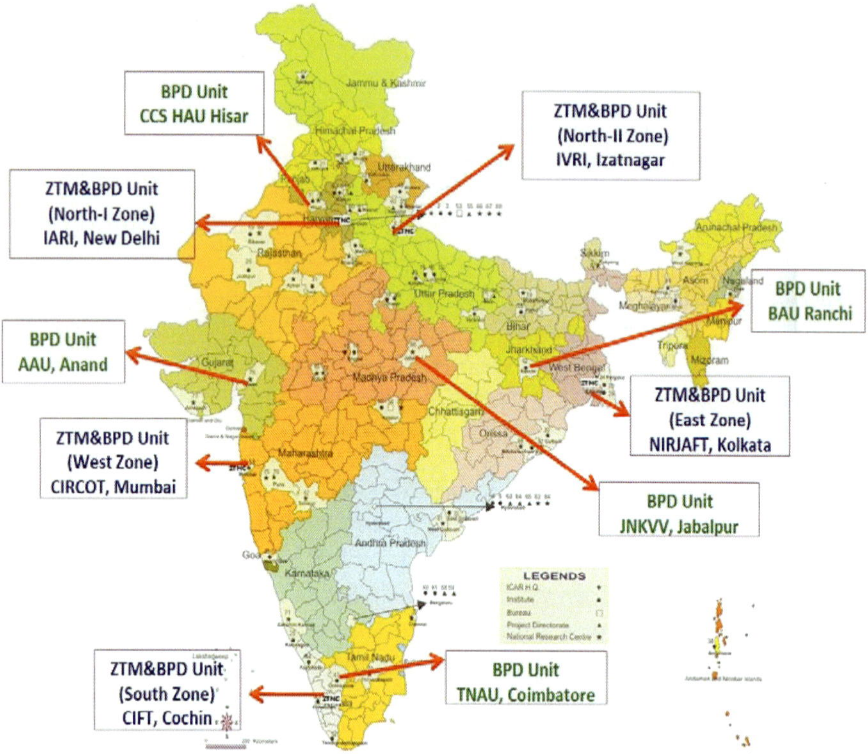

Action Points

Capacity Building: The "Intellectual Property Management and Transfer/Commercialisation of Agricultural Technology Scheme" broadly aims at up-scaling the IP management and technology transfer/commercialization activities in the ICAR set up. The ICAR institutes, either on their own or through suitable interaction with other institutes and ZTM-BPD Units at the zonal level, built capacity with respect to patent/IP search, IPR compatible record keeping, drafting MoUs, patent documents, licence agreements, confidentiality agreements, enhancing negotiation skills, patent/IP/market watch, etc. Effective IP management, commercialization (wherever applicable), and database management of intellectual assets constitutes important indicators of the performance under the scheme project.

Common licensing strategies: ICAR recognizes that TRIPS compatible IPR laws in India and in other member countries are important for management of agricultural research results. Once protected, these IPR enabled ICAR technologies, by way of licensing, could be transferred to end users through private, co-operative, non-governmental and public channels. Licensing could be for commercial use or for research or both. Application of incentives and benefit sharing with scientists/innovators and human resource development in ICAR improved the overall research environment and provided impetus for greater creativity and knowledge generation. Licensing of IPR enabled ICAR technologies encompassed out-licensing. The framework for licensing was developed/refined/evolved by ATMC/ZTMCs/ITMUs. Appropriate joint commercialization agreements were also entered into.

Valuation of technologies : Broadly, the worth of an IPR enabled technology will be derived from the likely benefits that may accrue to its end-users. The worth can be best determined on the judgment of technical experts, producers of technology and business managers. There is no standard method or formula for assessing the worth of a technology. Costs and pricing in ICAR will be determined on a case-to-case basis. The ICAR institutions will determine the licence fee and royalty and/or sale price of its IPR enabled technologies either on a fixed basis, through negotiations with the licensee, or through an open bidding process as appropriate. Expert opinion and judgment viewpoint together with the following points will be considered in determining the price/licence fee:

1. Cost of IPR protection and maintenance.
2. Cost of production and handling.
3. Other institutional costs as appropriate.

Pooling of technologies : ICAR recognizes that various NARS partners and related public research organizations in the country may be in possession of IPs that could be useful as resource/tool to enhance the overall research output. It may be cumbersome and expensive to procure and use such IP resources under the IPR regime, if each partner of a network or a consortium has to enter into separate legal agreements with each of the other partners. Therefore, ICAR has established a cost-effective and transparent mechanism of pooling IP resources for research in the NARS and by other public sector organizations in the country. ICAR has contributed to the development of a multilateral system of pooling and utilization of IP resources by interested partners.

Technology Transfer/Commercialization : ICAR transfers its IPR enabled and other technologies under the new regime in conformity with national priorities. Case-specific decisions will be taken regarding which technologies will be placed in public domain for open access and which others will be commercialized through non-exclusive or exclusive licences. All decisions on the mode of technology transfer will be preceded by the filing and recording of applications for protection of its IP. The IPR enabled technologies will be transferred for commercial purposes with suitable understanding agreement or contracts with the concerned parties. Specific terms of licensing can be negotiable. Commercialization will be undertaken either by ITMUs of the concerned institutions that have the requisite expertise and experience or the concerned ZTMCs/ATMC. Commercialization in foreign countries shall be undertaken by the ATMC.

Principles of Exploitation of Intellectual Property

ICAR encourages the development and exploitation of its intellectual property, to the benefit of the Indian Agricultural system, and as a part of its contribution to economic development. However, ICAR's prime role has been in the production and development of IP and the exploitation of IP takes place with the involvement of established and start-up companies, and other organisations. ICAR is open-minded about the mea1:

- Each ICAR R&D Institute has an obligation to make all reasonable efforts to protect commercially viable IP and derive an income from it. In such cases, the innovators will be appropriately recognised as contributors to that IP.
- Identification of potentially exploitable IP generated within the R&D Institute is the responsibility of the management of that Institute and ZTM-BPD Unit / ITMU. All technology innovator(s) and associated staff scientific staff have an obligation to identify any potentially exploitable IP and disclose to the concerned technology management unit (ZTM-BPD Unit/ITMU) operational at the respective Institute through an IP Disclosure Record; assistance with this should be made available from the ZTM-BPD Unit/ITMU.
- In the case of IP with potential for commercialisation, keep that IP confidential and ensure it is assessed and appropriately handled and protected before it is published or publicly disclosed

- Retain sufficient records, materials and data to justify the outcomes of research, and to assist in valuation and costing of the technology.
- Following receipt of an IP disclosure, evaluate whether to proceed with protection and/or commercialisation of that IP, providing preliminary advice on the disclosure within 1 month. If a decision is made to protect and/or commercialise the IP, determine and implement the best method by which the IP is commercialised on a case-by-case basis, in consultation with the innovators of that IP.
- Exploitation and commercialisation of IP might take the form of licensing, assignment or sale, to existing or newly-formed entities. This exploitation should not damage the Institute's core activities of research and development.
- Any interactions between private companies must strictly adhere to ICAR policies. Additionally, in general, such companies should not operate from Institute premises, except under a specific licence to occupy, which would normally be in a defined business incubator space.

Protection of Intellectual Property

- IP that is potentially exploitable should be identified and recorded using the IP Disclosure Record. The Institute management should strictly ensure the unprompted disclosures from staff and research students. In doing so, all parties will be cognisant of not damaging the Institute's core research and development activities.
- All identified IP will be reviewed by the ZITMC/ITMC, to consider its suitability and marketability. The Panel will take into account the wider strategies and policies of the Institute in determining the exploitation route. Protection of the IP will be determined and funded by the ZTM-BPD Unit/ITMU through the IP&TM Division.
- The Institute may make use of specialist external organisations, to take forward specific cases, where they bring detailed knowledge of the technology and/or relevant market. Interactions with such organisations will be managed by ZTM-BPD Unit/ITMU.

- The timescale for exploitation of any given set of IP depends significantly on market conditions as well as the state of development of the IP. The discussions between ZTM-BPD Unit /ITMU and creator(s)/innovator(s) should include agreement of a reasonable initial time-table, with regular review points. All parties involved have responsibilities in achieving successful exploitation, and need to be aware that circumstances will change as the process develops.
- Where an innovator is to undertake an involvement with a licensee or other external organisations, it must be governed by a suitable agreement between the company and the Institute, e.g. for consultancy, services, etc.

Distribution of Income from Exploitation

In general, the Institute will share the net proceeds of IP exploitation with the relevant staff after recovering its costs. The form of income can be as equity, dividends, royalty payments, or share of surpluses. Where IP has been sold, assigned or licensed, the Institute would normally expect to receive milestone and/or royalty payments. These will be distributed net of any relevant costs, such as legal bills, patent costs, and other protection and exploitation costs made as part of the development process. The distribution will also be net of any agreed revenue sharing with the funders of the original research. The information in the IP Disclosure Record will be used to define the creators of the IP, and their share of the proceeds. The net income from milestone and royalty payments to the Institute will be distributed on the basis of the prevailing rates, as set out in ICAR policy, which may be changed from time to time. This distribution will apply to each new agreement for any given item of IP. Renewals or extensions of existing licences will not be viewed as new agreements; i.e. where there is no substantive change to the subject, field, or territory.

Valuing IP Resources

Sometimes, all of the IP resources of the institution need to be valued. Valuing these resources can provide a value for the whole ICAR, including its intellectual property and physical resources, or it can reveal the input the Institute is investing into developing a product, excluding the technology that is being negotiated. One complication in these calculations is the need to value non-formal intellectual property: the know-how, experience, and expertise that reside in the Institute, and in its employees, and that may not be protected by patents and trademarks. Institutions that consider only the value of formal intellectual property stand to lose from overlooking this form of intellectual property. The primary reason for valuing IP is to maximise its value and therefore, the value of the owner organisation through optimum management decisions. There are various scenarios where valuation is required and needed, some examples are:

- **IP Exploitation :** IP is a fundamental component of an R&D Institute. An accurate valuation is required for exploitation of IP through business incubation/technology transfer/commercialisation activities. In such transactions, each party will need to know the value of IP assets being bought as part of the company.
- **Sale and license transactions:** Before an Institute commercialise its IP, it must be aware of its worth. Likewise, when negotiating a license contract, both parties must be clear about the values involved. Often, a due diligence report is required outlining the details of the IP being licensed.
- **Mobilisation of funds to Companies for business development (bank loans, venture capital, and investment) :** Properly valued IP would help the clients acquiring the IP of ICAR, to mobilise finance for the development plans of their companies. When making decisions about possible investment and associated risks, these organisations must be clear about the value and commercial viability of the IP belonging to the benefactor and often the reason for investment.
- **Taxation planning and compliance :** For legal entities, knowing the value of their IP is important for possible tax deductions and tax compliance.
- **Litigation support and dispute resolution :** Accurate IP appraisal is required in the event of IP rights infringement or breach of contract.
- **Internal management :** In the scientific enterprising drive of ICAR, successful exploitation of IP has great importance. IP exploitation and creation of business strategies requires effective management internally within the Institute. Research, development, legal, industrial protection application and

commercialisation decisions involve high but measurable levels of risk. IP valuation facilitates cost effective decision-making and helps to understand and deal with the risks involved.

Facilitation of IP and Technology Management at the Zonal level

The middle tier of IP and technology management mechanism implemented in ICAR, consists of 5 ZTM-BPD Units, which has the responsibility to cater to the individual and collective needs of various R&D institutions in its catchment area. The technological assets of these Institutions include a number of high yielding and resilient crop varieties, animal and poultry breeds and strains, packages of improved crop and animal husbandry practices, natural resource management technologies, improved tools, equipment and farm machinery, improved dairy, poultry and fisheries technologies, post harvest technology, computer software and data sets, and several other processes and products of agriculture and the allied sectors. As these institutions play a leading role in spearheading the scientific and technological development in India, the issues related to intellectual property rights are becoming increasingly important within the ICAR system. This is particularly so with respect to the current move by most institutes to generate income through consultancy, technology transfer and commercialization of innovations, inventions and research findings.

The role of ZTM-BPD Unit includes :

1. Securing and safeguarding IPRs of south zone institutes.
2. Effective IP management through collaboration by facilitating and reporting the relevant activities of member institutes.
3. To act as a Hub for member institutes for providing in incubator operations and protection & management of intellectual properties generated in the system, and its commercialization/ technology transfer for the public benefit.
4. To help entrepreneurs to commercialize business ideas utilizing the R&D back up of the institutes by providing value added business support services.
5. To develop and implement new services, organizational transformation and create a corporate culture that can foster entrepreneurship, and empower Agricultural Scientists for stimulating business innovation.

6. To popularize the ICAR technological assets in agriculture and allied sectors on an international level and build intensive partnerships with the private industry and entrepreneurs for ensuring a competitive presence.
7. To impart training and capacity building programme for, creating prospective entrepreneurs and generating value added manpower to compete effectively, create jobs, open up new economic opportunities for small and medium scale industries.

❑❑❑

Chapter 14

Mechanisms of Intellectual Property Management in ICAR

Atul P Kolte, A Dhali, Letha Devi G and N K S Gowda

The technological assets of ICAR include a number of high yielding and resilient crop varieties, animal, poultry breeds and fish strains, package of practices for improved agriculture and animal rearing, improved tools, equipment and farm machinery, improved dairy, poultry and fisheries technologies, post harvest technology, computer software and data sets and several other processes and products of agriculture and the allied sectors. Agricultural science has been the engine of growth and led to quantum jumps in the country's productivity in the past. ICAR technologies in farmers' fields and backyards have increased agricultural output and farm incomes. These technology packages have contributed to the green, white and blue revolutions. It is amply clear that ICAR must continue to produce significant research output to enable further enhancement of agricultural productivity and help to meet the future needs.

Protecting or patenting research output in agriculture was not customary in India and other developing countries prior to the establishment of WTO in 1995. Rarely any ICAR technologies were patented or commercialized. ICAR placed all developed technologies in the public domain for access by all. The TRIPS Agreement has led to the evolution of IPR regimes in WTO member countries. The Indian IPR laws are also made TRIPS complaint. ICAR recognizes that TRIPS compatible IPR laws in India and in other member countries are important for management of agricultural research results. Once protected, these IPR enabled ICAR technologies, by way of licensing, could be transferred to end-users through private, co-operative, non-governmental and public channels. Licensing could be for commercial use or for research or both. Application of incentives and benefit sharing with scientists/innovators and human resource development in ICAR would improve the overall research environment and provide impetus for greater creativity and knowledge generation.

Patents are looked as property and come with legal rights exclusive to the owner for use, license and sell as any other property. Today, every organization is safeguarding their inventions with the patent tool. The benefits of patenting an invention are as follows:

1. The patent holder has exclusive right to use the invention for manufacturing and preventing others to use the invention without permission of the owner.
2. The commercial rights of patent holder can earn money by selling or licensing the invention to others.
3. Patent laws provide protection to the invention for 20 years.
4. Patents are partially responsible for advancement of medical science, biotechnology, pharmaceutical research, etc.
5. Patents are important assets in universities and education institutes opens up opportunities for knowledge sharing with others (universities, Institutes, industries, etc.)
6. The above benefits rewarding the inventors motivate the researchers to lead the studies to usable output for society.

ICAR recognizes that intellectual assets in agricultural research, education and services and technology transfer have assumed increasing importance under the current IPR regime. Bridging the gap between invention and commercialization is very important. Technology transfer and diffusion play a crucial role in enhancing the agricultural production and development. Prompt and efficient availability of ICAR technologies and know-how to farmers is the key issue. A significant presence of the private sector is additionally helpful in extending the ICAR's technology and intellectual services.

The interaction between scientists and industry is also crucial for successful commercialization of inventions. Institution owned patents could be used as a collaborative tool with industry. The entrepreneur often does not have the knowledge and insight about the basic research that a scientist carry out at lab level and the scientist on the other hand, not aware about the up-scaling requirements, commercial production and marketing methodologies for the invention. By retaining patent rights and licensing patents to industry, Institutes can create a very strong platform, where both the scientists and entrepreneurs can interact and understand the perspective. This interaction and joint effort can bring easy commercialization of the inventions. Incentive to invent is very important for any scientist. Scientists will be happy to know that their invention is being used commercially. This boosts the moral and person starts working with double energy. This will lead to efficient work and more inventions, which will lead to further patents and profit to the organizations.

ICAR has taken several steps towards developing and strengthening the decentralized three-tier IP management system, which came into effect on 2nd October, 2006. The salient features of the initiative are as follows:

- Executive powers were delegated to the Directors of all ICAR Institutes for signing IPR applications and executing license agreements as authorized signatories of Secretary, ICAR. This was aimed to improve the efficiency and technology transfer process.
- ICAR scientists and concerned staff engaged in IP management at its institutes have been sensitized through various workshops, training programs and regional meetings.
- Five national Institutes under ICAR (CIFT, Cochin; CIRCOT, Mumbai; IARI, New Delhi; IVRI, Izatnagar and NIRJAFT, Kolkata) have been identified to establish Zonal technology management centers and were made responsible for facilitating and reporting the relevant activities and outputs of the ITMUs of all ICAR institutes in the respective zones.

Central IP Management

At the top tier, ICAR facilitates training and capacity building, besides general facilitation of IP&TM, The highest-level decision-making is under the guidance and approval of Director General and routine matters are processed at the level of ADG (IP & TM) at ICAR HQ. The Central Technology Management Committee (CTMC) provides the decision support for IT&TM and BPD requirements. The governing body of ICAR has empowered DG, ICAR to implement and also modify the overall IPM&TTC guidelines, if necessary. The general purpose of these initiatives is to promote smooth development of the ICAR's IP&TM domain and the promotion of IPR-compatible research culture. However, the policy changes with major consequences are needed, the matter shall be referred to the Governing Body of ICAR.

Zonal Level IP Management

The five national institutes at the zonal level already established their respective Zonal Technology Management Centers (ZTMCs). These centers are also empowered with Business Planning and Development (BPD) units under NAIP support. Thus, the zonal level centers are now re-designated as ZTM&BPD units for facilitating the ICAR's IP management and business development activities in all its institutes located in their respective areas. The decision support system to ZTM&BPD units is provided through the Zonal Institute Technology Management Committees (ZITMCs). The directors of all these zonal ICAR institutes already have powers to sign the IP management related documents on the behalf of Council. The ZITMC has the following composition: the committee is chaired by Director of the Zonal Institute and co-chaired by Director of another Institute in the zone. The members include Director/representatives of another ICAR institutes whose matters are included in the agenda items, Chief or senior finance officer, Technical experts (ICAR scientists), Representative from ICAR, expert members one each in Business management/commerce and law. The member secretary to the committee is in-charge ZTM&BPD unit.

The functions of the ZTMU are as follows:

- The units must comply with the ownership and licensing right issues as the the overall ICAR guidelines. ICAR has to be represented as the sole or joint owner of the IP generated in the set up or the zonal and other institutes must represent themselves as constituents of ICAR and execute agreements only as per the delegation of powers.
- The units should ensure that concerned inventors/innovators have assigned the IP generated by them to ICAR and, in lieu, they have been recorded as true and first inventors/innovators/authors in the patent/IPR applications.
- The units should provide all possible needful assistance or guidance to the institutes in their respective zones.

- The zonal institutes need to collect and collate traceable records of the IP management activities of all ICAR institutes falling in their respective zones, maintain the zonal IP database and they must periodically update the central IP database in consultation with the IPR unit of ICAR.

Institute Level IP Management

In the three-tier IP management in ICAR, various primary activities, such as disclosure of IP generated, patents/IPR, prior art search, patent/IPR application writing, filing of applications at the concerned granting authorities, pre-grant and post-grant follow up, offering/advertising for commercialization, commercializing, license fee/royalty collection, benefit sharing are to be handled in the institute level technology management units (ITMU).

The mechanism for the establishment and functioning of ITMU and ITMC are as follows: The first pre-requisite for organized IP&TM at all ICAR institutes is that they must establish an ITMU and constitute an ITMC at the institute level. The ITMC will have Chairman (Director of the Institute) and members including Head of Division, Technical experts, IPR expert from ICAR institute in the zone, Member Secretary of IRC and other experts or officers of the institute as approved by the Director. The Officer-in-Charge ITMU serves as member secretary of the committee. The composition of the ITMU should be duly notified, circulated and documented in the IP&TM database. Other institutions like AICRPs, network projects, KVKs which are located outside ICAR Institutes should not take their individual decisions, but they must undertake their IP&TM activities with the concurrence of the respective ZTM&BPD unit.

The guidelines and functions of ITMUs are as follows:

- Decision support system to ITMUs is provided through the Institute Technology Management Committee (ITMC).
- Directors of all ICAR institutes have been delegated with the necessary powers of the Secretary, ICAR to sign the IP management related documents on the behalf of Council.
- ICAR is sole or joint owner of the IP generated in the Institutes. The institutes must not claim IPR in their own name. The inventors/innovators are in lieu provided the right to be the true

authors of the IP invention and are also entitled for benefit sharing from the net proceeds of commercialization of their inventions as per the guidelines.

- The institutes must present themselves as constituents of ICAR and execute the agreements with other parties on behalf of ICAR only.
- If the concerned institute lacks experience or confidence for seeking protection, commercialization, managing particular IP or business licensing deal, the respective ZTMC may be approached for needful assistance.
- The institutes are responsible for maintaining all traceable records of their IP management activities, and maintain their IP database. They must send copy of their IP database for updating at respective zonal units and also to ICAR.

Table 1: Coverage of ICAR Institutes under different zones and zonal Technology Management and Business Planning and Development Units (ZTM&BPD)

Zone	Institutes
Zone North-I	**Indian Agricultural Research Institute, New Delhi ZTM&BPD Unit**
	CAZRI Jodhpur, CIAH Bikaner, CIPHET Ludhiana, CITH Rangreth (JK), CPRI Shimla, CSWSRTI Dehradun, CSSRI Karnal, DMR New Delhi, DWR Karnal, IARI New Delhi, IASRI New Delhi, IIVR Varanasi, NBPGR New Delhi, NCAP New Delhi, NCIPM New Delhi, DMR Solan, DRM Bharatpur, NRC-PB New Delhi, NRC-SS Ajmer, PDCS Meerut, VPKAS Almora
Zone North-II	**Indian Veterinary Research Institute, Izatnagar ZTM&BPD Unit**
	CARI Izatnagar, CIRB Hisar, CIRG Makhdoom, CISTH Lucknow, CSWRI Avikanagar, DSR Mau Nath Bhanjan (UP), IGFRI Jhansi, IIPR Kanpur, IISR Lucknow, IVRI Izatnagar, NBAIM Mau Nath Bhanjan (UP), NBAGR Karnal, NBFGR Lucknow, NDRI Karnal, NRC-AF Jhansi, NRC-E Hisar, NRC-C Bikaner, D-CF Bhimtal, PD-C Meerut, PD-FMD Mukteshwar
Zone South	**Central Institute of Fisheries Technology, Kochi ZTM&BPD Unit**
	CIBA Channai, CIFT Kochi, CMFRI Kochi, CPCRI Kasargod, CRIDA Hyderabad, CTRI Rajahmundary, CTCRI Thiruvananthapuram, DOR Hyderabad, DRR Hyderabad, IIHR Bangalore, IISR Kozhikode, NAARM Hyderabad, NIANP Bangalore, NRC-B Thiruchirapalli, D-CR Puttur, D-OPR Pedavegi, NRC-M Hyderabad, DSR Hyderabad, NBAII Bangalore, NIVEDI Bangalore, PD-P Hyderabad, SBI Coimbatore
Zone East	**National Institute of Research on Jute and Allied Fibres Technology, Kolkata ZTM&BPD**
	CARI Port Blair, CIFRI Barrackpore, CIFA Bhubaneswar, CRIJAF Barrackpore, CRRI Cuttack, ICAR -RC-ER Patna, ICAR-RC-NHER Barapani, IINRG Ranchi, NIRKAFT Kolkata, NRC -L Muzzaffarpur, NRC -O Pakyang (Sikkim), DRWA Bhubaneshwar, NRC-M Dimapur, NRC-P Guwahati, NRC-Y Dirang, DWM Bhubaneshwar
Zone West	**Central Institute for Research on Cotton Technology, Mumbai ZTM&BPD**
	CICR Nagpur, CIRCOT Mumbai, CIAE Bhopal, CIFE Mumbai, ICAR -RC-Goa, IISS Bhopal, NBSSLUP Nagpur, NRC -C Nagpur, NRC-G Pune, D-Groundnut Junagarh, D-MAP Anand, D-OG Pune, D-Soybean Indore, NRC-WS Jabalpur, NRC-P Solapur, SPUC Bhopal

References

ICAR. 2006. ICAR Guidelines for Intellectual Property Management and Technology Transfer/Commercialization. Indian Council of Agricultural Research, New Delhi.

ICAR. 2010. Institutional mechanism and good practices for IP&TM-A check list. Information brochure 2010/1. Indian Council of Agricultural Research, New Delhi, Pages 1-12.

Mauria S, Kochhar S and Saxena S. 2009. ICAR ZTM&BPD perspective: zonal technology management and business planning and development meetings-cum-workshops 2009-10. Information brochure 2009/1; ICAR, New Delhi, Pages 1-26.

❑❑❑

Chapter 15

Organizational Designs for Enhanced Patent Productivity

Mary Mathew and Urmi Chattopadhyay

Introduction

Intellectual Property (IP) today is seen as a principal source of shareholder wealth and competitive advantage. The IP of an organization is a wealth pusher. Stock values of organizations who have a well-designed strategy for IP, and accumulation of IP within their organization, are likely to be robust, and growth bound. Thus, the knowledge economy is giving rise to a new type of business market, the "Intellectual Property market". Here in lies one of the great corporate challenges: figuring out how to unlock the hidden power of patents through appropriate organizational design. Organization design (Jones and Mathew, 2010) refers to designing the components of an organization in order to help it perform better. The components being structure, tasks, processes, culture, technology support, human behaviours, goal clarity to name a few. This paper looks at organization design for patent productivity. This puts into focus the role of the R&D departments in organizations.

Unlocking IP within an organization requires a focused drive towards result oriented R&D. The encouraging aspect of R&D is that, while the R&D project group makes attempts to reach the result along the way, the group may realize other unexpected results, and this seems to be a common experience. Thus, R&D is characterized by planned and unplanned results. Both sources of results are required to unlock IP within the organization. An invention orientation is

---

Adapted from Mathew, M., and Urmi, C., Entrepreneurial R&D: An Organizational Design for Enhanced Patent Productivity, **Portland International Conference of Management of Engineering and Technology (PICMET),** Portland, Oregon, USA, July, 2001

required to make this happen. An invention orientation within R&D, is one in which the person is seeking *novelty* and *utility* in results continuously. The opportunity orientation is also high, and non-judgmental.

Researchers and consultants (Ransley and Rogers, 1994) have studied best practices for value added R&D with a patent focus. Literature in R&D Management correlates organizational design variables including, structural variables, especially the role of cross functional teams in R&D (Glenn, 1985 and Purdon, 1996), psychological dimensions (Ernst, *et al.*, 2000), and HR correlates (Koning, 1993, Gupta and Arvind, 1993). Studies that have focused on patents as a main output variable have done so by looking at the number of patents generated. While number of patents are a popular count, the how and why of patents within the organization, requires deeper analysis.

The organizational design correlates that maximize patent productivity or are the push factors provide the ground for re-designing organizations to maximize IP. A metaphor from rain forests is worth mentioning here. The rain forests are constantly sprouting vegetation, shrubs, mushrooms, and other varieties of living beings. The atmosphere is conducive to this thick and dense growth. An organizational design suitable for this dense growth of IP within it, must have an organizational design which is an IP enabler. Two approaches to understanding the push factors of IP are described, first, is the understanding an organization has of its patent productivity measures itself (performance measures), and second, is the diagnosis of whether the internal organizational structure, task, processes, that maximize IP, are in place or out of place.

Patent Productivity

Patent productivity can be referred to as the drive to patent and protect scientific and technical knowledge, that can be defined as novel, unobvious and having utility. It is a performance measurement of patent outputs. Organizations that are "patent pushers", (such as the ones seen in Table 1 are those that are high on patent output. Patent productivity can be looked at from two levels of analysis: (a) organization as a whole (b) patent specific characteristics. One measure of output that is popular for organization level comparative analysis, is the number of patents generated annually. The list of the top twenty organizations that are patent pushers indicates that IBM is the leader, and it has been so for the last five years, with 1742 patents in 1998 (The Patent Intelligence and Technical Report from the IFI Claims Patents Service). IBM retains this position in 2008 and 2013 also. These statistics only indicate USPTO filing and world statistics showing a rising curve for China in patent filings.

Organizational comparison of patent outputs, both between and within industry is a good indicator of the competitive advantage of the organization. In this sense, IBM has an apparent edge that over the others. So, do the top organizations in the world market. Organizations attempting to understanding their position by number of patents produced are on the right track towards sustained growth and shareholder wealth. The overall supremacy they feel will be high. Organizations that can maintain this high, will bounce back when faced with a problem. The leading organizations that file at USPTO are shown in Table 1.

Table 1: Showing the top filers in USPTO

1998		2008		2013	
IBM	1742	IBM	4169	IBM	6788
Canon (JP)	1381	Samsung (KR)	3502	Samsung (KR)	4652
NEC Corp (JP)	1101	Canon (JP)	2107	Canon (JP)	3825
Motorola	1065	Microsoft Corp	2026	Sony (JP	3073
Mitsubishi (JP)	925	Intel	1772	Microsoft Corp	2659
Hitachi (JP)	922	Toshiba (JP)	1575	Panasonic Corp	2582

The Table shows variations in organization patent productivity. This implies variations in organization designs that enable this productivity. The apparent edge is with the growing microelectronics and ICT industry, with large numbers of software and electronic goods patents. A mere look at patent counts, as a productivity measure may be inadequate. The numbers require further analysis on an extraction metric in a patent specific way. Patents vary in their commercial potential. They can be classified into two poles in a linear continuum, namely: (a) patents that were filed for no apparent reason, (b) patents that enhance the knowledge base, through to, (c) patents that maximize commercial returns. Patents that are filed for no apparent reason are those that slipped past the "to file or not" query during a patent strategy assessment that goes on in-house with the IP department of the organization. Repeated such filings, will make the organization consider whether there is need to carry on with the R&D activity associated with those patents. On the other hand, patents that are rather fuzzy to the lay eye, are those that fall into category (b) and they are patents that enhance the knowledge base of the organization. These contribute to the body of societal knowledge for that patent family. They are important as they incrementally build upto an eventual patent, over time. Such a patent will in the future, maximize commercial returns. Hence, they belong to that part of the continuum of no immediate commercial returns. On the other hand, patents that maximize commercial returns are those that have immediate revenues in the forms of royalty, licensing or technology sell outs. The crown jewel patents are those that bring extraordinary returns, far surpassing their inputs. They fall into the side of the continuum that is having high commercial returns. Figure 1 below graphically represents this theoretical extraction metric:

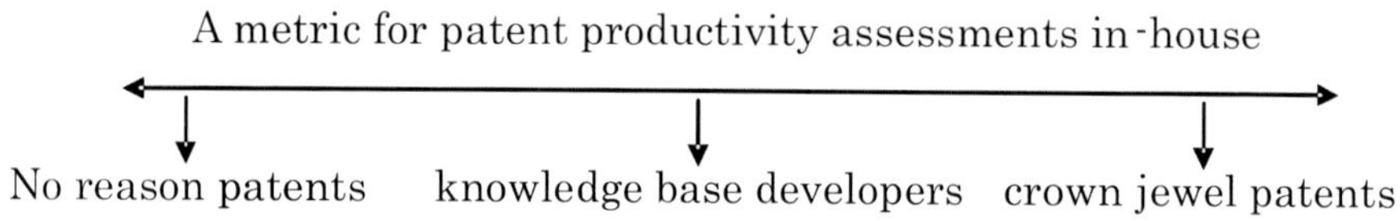

Fig. 1: Showing the extraction metric for in-house patents

Analysis of patents within the organization, classifying them in this manner and then doing a within and between organization level comparative analysis, will provide value addition to the patent productivity assessment of the organization from time to time. Patent productivity also requires other indices which are intra organizational. These are :

a) Annual Patent Expenditure

Annual Patent Revenue

Where Annual Patent Expenditure includes costs specifically related to the patenting process, namely:

Filing Cost: *Filing of the patent both at the national level and abroad involves a cost. The cost depends on the number of countries, where it is filed (foreign filings) and the cost of filing in those countries.*

Maintenance Cost: *Fees for maintaining a patent mount with time. Broader filing involve higher costs to maintain with the renewal fee.*

Translation Cost: *Filing of a patent application abroad involves a cost of translating it in the national language of that country, precise translations go at a premium rate.*

Prior Art Analysis Cost: *The provision of patent search facilities in the organization requires expenses in the form of getting online databases from national and international sources, CD ROMs and generating meaning and knowledge out of the output. The cost referred to here, is of generating knowledge data of past efforts.*

Prosecution Cost: *This is for paying representatives to amend and argue the case. The whole process involves interaction with attorneys and patent agents involving huge costs.*

Annual Patent Revenue **includes all forms of returns from patents, namely immediate royalty, royalty down the patent tree, licensing, equity, sale of patents or technology sell outs related to that patent.**

b) Patent Budget

R&D Budget

The patent budget refers to the financial amount allocated for patent expenditures, annually. R&D budget, on the other hand, refers to the total R&D budget annually. The patent budget allocation indicates the importance given by the organization to the patenting process. Lower fractions of the ratio, indicate poor planning for the patent process and high planning for the input process, namely R&D. This calculation is also possible at the R&D project/patent levels too.

This approach requires Finance and Accounts working with R&D teams and going beyond the count of patents in order to measure productivity.

Internal Practices: Patent Pushing vs Patenting Trying Organizations

The interesting difference in patent output amongst the top organizations in Table 1, motivates internal within organization analysis. The internal organizational environment may contribute to patent pushing and obtaining or patent trying and not getting an awarded patent. Table 2 shows a comparison of the possible traditional and new organizational design variables. The new variables are focused on IP generation.

Two of top filing organizations were visited to understand the push factors that may facilitate patent output at the organizational level. Inventor centered interviews were conducted in Bangalore city. Both organizations were from the IT industry, and giants in IC designs and semi-conductor R&D. An interview schedule was used to ask questions related to the above variables and the flow of the organizational patent process.

Table 2: Showing a comparison of the traditional organizational design variables and the patent pushing variables:

Traditional variables	Patent pushing
Organizational structure, namely cross functional team structures, and decentralization of R&D	*Team task roles defined around idea data basing, patent searching skills, infringement analysis, patent writing and other IP activities* *Team empowerment regarding knowledge about IP, maximizing decentralization*
Monitoring and control of projects through project review meetings	*IP maturity review at project reviews: off loaded to IP know-how empowered teams, stage pushed lower, maximizing control at lowest end*
R&D Process flows	*IP Process flows superimposed on R&D process flow*
HR initiatives in rewards and definition of performance parameters	*HR initiatives in recognition and reward for IP generating inventors, and performance parameters benchmarked on an IP extraction metric*
Collaborative R&D based on common or complementary know how	*External learning through collaborative R&D for IP synergy, and markets strung around common inventor themes*
Cost centric, budget driven R&D	*Revenue centric R&D*
Combined looked at output variables, papers, citations, patents, financial variables	*Extraction metric related assessment of IP as the main output variable for R&D or new product development*

Excerpts from the case are narrated below:

Case 1:

This organization intends to be the leader in building a broadband and mobile Internet infrastructure. It provides the systems, software, silicon and services for what is the largest network build-out in history. The company is focused on a triple play of optical, data and wireless networking technologies with the software and services to support them. The organizational process of patenting were discussed. The R&D head of the microelectronics division of the organization narrated his experience on arriving at the patent. "In the process of the progress of one project, two problems were faced by my team. We reported these problems to the local management in a project review meeting. After a brief discussion the management asked us to work in the direction of finding the solution to these problems. The team worked towards trouble shooting the problem without an idea of ending up in a patent. A solution was found to the problem. Even at this stage, we did not realize that it was an innovative solution, which can lead to a patent. However, the local management felt that the solutions were novel and patentable during the second project review meeting. The papers were immediately sent to the Intellectual Property Division in USA and there, they filed the case for a patent. The attorneys were involved at this stage and we had to send in write-up about the problems and the solution obtained. The Patent Search was conducted by the IP Division in USA."

Case 2

The second organization was a global Semi-conductor Vendor specializing in DSP and Analogs. It is a leading designer and supplier of digital signal processing and analog technologies, the engines driving the digitization of electronics. They are moving fast to drive the Internet era forward with semiconductor solutions for large markets such as digital wireless and broadband access. The R&D head of this organization was interviewed regarding the process of patenting that is followed there. The person interviewed had been granted a patent recently and he described the process flow that he followed. He elaborated that the idea came from the office in USA, in response to a customer need. The concerned person started working on this. He was aware of that he was doing something innovative and novel but the idea of getting a patent was not there. The R&D involved with that need, led to a new and novel solution, which was sent to the IP cell in USA for filing. The procedures thereafter were handled by them. At no stage of the process, the scientists were involved in prior art search or interactions with attorneys or patent agents.

Looking at the variables from Table 2, comparisons can be drawn regarding the internal organizational analysis. Both organizations use cross functional teams and R&D is more or less decentralized. However, the task roles of are IP pushers were unclear and undefined. Both organizations did not have IP oriented task roles. The patent searching exercise and the infringement analysis were centralized activities conducted at the parent organization at USA, not in India. The team, thus, was not empowered with IP oriented know-how to maximize patent output. Decentralization of R&D will reach its fullest form if and only if, teams are empowered to realize IP arrivals within their task definitions itself.

Monitoring and controlling is a common practice in both organizations, this is done through periodic project reviews. It is at these reviews that the IP awareness creeps in. However, the team per se is not equipped with the knowledge required to be IP aware within the task. Thus, team monitoring of IP output is something that will take the organizations a longer way with IP generation. Task roles can be re-design as intrinsically IP oriented and having potential for its happening.

The R&D process flow starts from idea generation through to award and practice of the patent. The process of idea generation was either client centered and having roots from the parent organization as in case 2, or problem solving centered and local, as in case 1, which is also client driven. Thereafter, the important decisions appear to center around the project reviews, and the rest of the time the team is working on the software development, mainly coding, unaware of the IP generation within their work. The processes takes a turn, when there the senior managers during the project meeting realize that there is possibility for IP, and re-directs this to the IP department at USA. In turn, they forward the request to process of filing for a patent. The IP process must identify IP milestones and must check these milestones out periodically (Mathew, 1996).

Conclusion

In conclusion the process described above in both cases, indicates the tasks by which a patent emerges. Multinational organizations with R&D bases in other parts of the globe (India, for instance) other than the parent location, handle their IP process in a semi-centralized design. When IP is centralized, it means that the IP department looks after the entire IP tasks and processes and the department has its own structure which will interface with R&D, finance, marketing, strategy and top management. The decentralized design ensures that R&D scientists execute IP related tasks with a minimal presence of a full fledged IP department.

The idea generation source in Case 1 was local, and the idea generation source for Case 2 was from the parent location. The patent search, which is conducted to solidly understand the prior art that went into previous patents, is a process that was done much later than expected in both cases. Further in both cases, this process was conducted at the parent location. The filing and writing of the technical specifications for the patenting procedure was also handled in a centralized manner at the parent location. The awarding will also be done by the parent location and the inventors names will be mentioned on the patent, but owned by the organization.

Although these are performing organizations at their Indian location their current organizational designs fail to empower their teams with IP know-how, like queries on: How to recognize a patent when it comes? When to patent? What to patent? And Where to patent? Currently, these processes are semi-centralized and decisions are located at the parent location, away from local R&D centers. The case studies throw light on the fact that studying the number of patents alone will be inadequate and in-depth analysis is required based on these two approaches. Re-designing a conscious organization thereafter will enable the organization to get closer to the entrepreneurial R&D organizational design, which will be by default, patent productive.

References

Berkowitz, Leonard. "Getting the Most From Your Patents," Research. Technology Management, Mar-Apr 1993, pp 26-31.

Ghoshal, Sumantra and Barlett, Christopher. "Building the Entrepreneurial Corporation! New Organizational Processes, New Managerial Tasks," European Management Journal, Vol.13, No.2, June 1995, pp 139-155.

Gupta, Ashok K. and Singhal, Arvind. "Managing Human Resources for Innovation and Creativity, "Research. Technology Management, May-June 1993, pp 41-48.

Gwyne, Peter. "Managing Multidomestic R&D at ABB " Research. Technology Management, Jan-Feb 1985, pp 30-33.

Mathew, Mary "Benchmarking Patenting Procedures: A Conceptual Model of the Procedure and Experiences of the R&D Scientists", PICMET, 1996.

Purdon, William. "Increasing R&D Effectiveness: Researchers as Business People," Research .Technology Management, July-Aug, 1996, pp 48-56.

Ransley, Derek L. and Rogers, Jay L. "A Consensus on Best R&D Practices," Research. Technology Management, Mar-Apr. 1994, pp 19-26.

Shapiro, Amram R. "Responding to the Changing Patent System, " Research . Technology Management, Sept-Oct. 1990, pp 38-43.

Stevens, Greg A. and Burley, James. "3000 Raw Ideas=1 Commercial Success! ," Research .Technology Management, May-June 1997, pp 16-27.

Taylor,Glenn L,Snyder, Louis J, Dahnke,Keith F and Kuether, Gary. "Self Directed R&D Teams: What makes them Effective," Research. Technology Management, Nov-Dec. 1985, pp 19-23.

Jones, Gareth and Mathew, Mary, 2010, Organizational Theory, design and change (6th edition), Pearson, India.

❑❑❑

Chapter 16

NRDC - Transforming Innovative Research into Profitable Technology

N G Lakshminarayan

The accelerating spread of new technologies in a competitive environment world-wide makes it imperative that new ideas and technologies are identified at their source in the scientific world, carried forward by technological innovation and piloted into commercial production.

NRDC is India's premier service enterprise whose business is to be the identifier, the carrier and the pilot of technology transfer. For over fifty years, NRDC has played a key role in speeding the commercial applications of research and in effecting the transfer of technology from laboratory to enterprise. NRDC guides and assists the entrepreneur in executing his technological business plans.

NRDC works in close conjunction with over 200 national R & D laboratories and has licensed over 2000 technologies for commercial exploitation, of which nearly 1000 are in production with a current annual turnover of about Rs.12,000 million (Rs.12 billion).

Technologies licensed by NRDC cover areas, such as agriculture, horticulture, sericulture, food, agro-processing, bio-technology, chemicals, drugs and pharmaceuticals, metallurgy, electronics, instrumentation, building materials, manufacturing techniques and utility processes including pollution control.

Major Activities of NRDC

Commercial

Development and Commercialization of Technologies are being undertaken under the Programme for Development of Technologies for Commercialization (PDTC) with the objectives of:

- Technical evaluation of the innovations/technologies.
- Licensing of technologies to companies & end-user organizations.
- Providing grants for development of Technologies to R&D laboratories for setting up pilot plants to validate or scale-up laboratory processes prior to commercialization.
- Participation in equity of early stage ventures.
- Intellectual Property Rights (IPR) consultancy to R&D Institutes and Corporate.
- Export of Indian technologies/know-how and services.
- Execution of turnkey projects abroad based on indigenous technologies.

Promotional

Promotion and Commercialization of Inventions

- Awarding meritorious inventions.
- Assisting innovator & inventors in patenting and commercializing their inventions.
- Popularization of science highlighting inventions by bringing out magazine "Awishkar" (monthly) in Hindi.

Promotion & Development of Household & Rural Technology

- To identify, prove and demonstrate selected rural & household technologies.
- Assisting in commercialization of selected rural & household technologies.

Promotion of Export of Technology

- Projecting India as a source of technology.
- To prepare publicity material like catalogues, brochures, multimedia CD-ROMs giving information about the technologies suitable for export.
- To prepare feasibility reports and basic engineering design packages for the Information on Technology Transfer.
- Organizing training programs related to Technology Development and Transfer.

- Participating in exhibition, publishing periodicals, arranging audio-visuals to popularize indigenous technologies, both at home and abroad.

Disseminating Information

NRDC, as a service organization, constantly endeavours to encourage inventive talent in the country and to promote development and utilization of results of indigenous research with a view to accelerate the technological and economic self-reliance of the nation. To achieve these ends, NRDC brings out the following publications to promote its technologies by disseminating information through them:

- Under its invention Promotion Program brings out popular Science and Technology monthly magazine - Awishkar in Hindi. This magazine provides information on science and technology to entrepreneurs, scientists, technologists, industrialists, inventors, students and the public.
- NRDC Processes-A publication brought out periodically, contains lists of available processes classified based on various disciplines of technology.
- Pamphlets and booklets are also brought out on selected and successful NRDC technologies from time-to-time.

To create awareness on the range of technologies and services that are available from India, NRDC participates in exhibitions, seminars, workshops and entrepreneurship development programs both in India and abroad.

The Sources of NRDC's Technologies

India has an extensive network of about 2500 R&D institutions, mostly Government supported. These institutions cover virtually every branch and facet of research and technology ranging from paleo-botany to spacecraft. The institutions in this network are the primary sources of NRDC's portfolio of marketable technologies. In pursuance of its corporate goals, NRDC has forged links with, and between the scientific and industrial communities in India and abroad. Corporation has formal arrangements with many of the major R& D organizations in the country for the commercialization of know-how developed in their laboratories.

Technology Marketing Experience

NRDC has accumulated experience in matching its portfolio of source points of technology to their potential markets in about 4600 cases, both in India and abroad. Countries from South often find Indian technologies compatible to their needs. Their simplicity, smaller plant sizes, lower investment levels, ability to utilize local resources and employment generation capacity, make them "appropriate technology". As a result of NRDC's assistance to its licensees in identifying potential clients, providing marketing support and obtaining product certificates from authorities in different countries, where these are required, the NRDC licensees have been able to export their products to both the developing and developed countries.

NRDC IPR LINE - A fine line that is keeping you ahead in an increasingly competitive business world

NRDC is India's premier technology service enterprise whose business is to be the identifier, the carrier and pilot the technology transfer. For over five decades, NRDC has played a key role in spreading the commercial applications of research and in effecting the transfer of technology from laboratory to enterprise. By providing business critical information to companies and research institutes across the country, NRDC has maintained its position among the top information providers on IPR. Its analysis of technical data is second to none. NRDC's IPR solutions have met the needs of companies and, equally important, fulfilled the technological demands of more specialized organizations—ultimately to benefit the organizations of the future.

Our patent assistance will provide you with a comprehensive picture of technological innovations worldwide—and also provide critical advantage in highlighting the new opportunities, identifying competitors, and assisting R&D. Our back record position us as leaders in providing IPR Consultancy to individuals, SMEs, R & D institutions, Govt. funded laboratories, IITs, Universities, etc. We are committed to remain at the forefront for protection of technology innovations.

Intellectual Property Facilitation Centre (IPFC) at Regional Office, Bengaluru

The IPFC has been setup at Regional Office, Bengaluru jointly by Ministry of Micro, Small and Medium Enterprises, GoI and NRDC.

The objective of this Centre is to provide following intellectual property rights related services to MSME sector:

- Searching, mapping, etc. with respect to patents, trademarks and industrial designs.
- Basic information to file applications for grant of patent, trademarks, industrial designs and geographical indications.
- Guidance in filing IP applications with National/International agencies.
- Collaboration with potential clients for technology up-gradation, tie-ups, scale-ups, etc.
- Advise on legal remedies on infringement issues.

Support to Inventors

The Programme for Inspiring Inventors & Innovators (PIII) of NRDC endeavours to stimulate the spirit of inventivity and innovation not only among scientific and technical personnel and industrial workers but also amongst craftsmen, artisans and the emerging youth in the community. In order to harness their creative talent for the benefit of the nation, the IPP covers a wide range of activities, including:

- Prize awards for meritorious inventions.
- Assistance for patenting inventive ideas of R&D organizations and universities both in India and abroad.
- Assistance to inventors to commercialize their patented inventions.
- NRDC gives cash awards under three categories and certificates to inventors by way of recognition of meritorious work. The awards are announced every year for meritorious inventions. The cash awards are exempted from Income Tax. NRDC also helps inventors in arranging the commercial exploitation of their patented inventions.

Reaching out to Rural India

Rural India with its urgent need for technological inputs, provides a vast market for new agricultural processes and services.

NRDC promotes technologies that are particularly suited to local needs. Examples include: the agro-waste utilization paint derived from the latex of Euphorbia plants, the mud block making machine and the use of blue-green algae for nitrogen fixation and farm soil enrichment. Efforts are being made to reduce the drudgery of women in their daily lives, to provide literacy aids, safe drinking water, low cost housing, sanitation, better agricultural implements and post-harvest technologies, solar pumps, heaters, solar cookers and solar dryers, cement from paddy husk ash and to provide employment by the utilizations of farm wastes. NRDC has also set up 52 rural technology demonstration cum training centers in various parts of the country to give hands-on training and demonstrations to artisans and technicians living in rural areas.

Working Together

NRDC shall continue its efforts in strengthening the Industry-Academia partnerships of various organizations and academic institutions from India and abroad, for propagation of the country's technological capabilities in wide-ranging sectors and applications.

Summing-up, NRDC is your Window to:

- Information on the latest technologies developed around the world and particularly in India
- An objective assessment of the commercial potential of inventions
- National and international patenting
- Identification of potential manufacturers
- Negotiation of licensing agreements
- Legal back-up for royalty collection
- Share in license income and other unexpected gains
- IPR consultancy to firms, R&D Institute, etc.
- Awarding meritorious inventions.

❐❐❐